ANIMALS AND MAN
George Cansdale

The changing pattern of relationships between man and the rest of the animal kingdom, from prehistoric times to the present day.

Ploughing near Ambleside
Picture Post Library

Animals and Man

by GEORGE CANSDALE
Superintendent of the
London Zoological Gardens, 1948-1953

WHITE LION PUBLISHERS
London and New York

Copyright © George Cansdale, 1953

First published in the United Kingdom
by Hutchinson & Co. Ltd., 1953

This White Lion edition, 1974

SBN 85617 064 X

Printed in Great Britain by
Biddles Ltd., Guildford, Surrey,
for White Lion Publishers,
138 Park Lane, London W1Y 3DD

CONTENTS

	Page
Introduction	9

PART I

THE RESULTS OF COMPETITION AND CONFLICT

1. In Great Britain   13
2. In Africa   26
3. In North America   36
4. In other lands and at sea   41

PART II

THE RESULTS OF INTRODUCING FOREIGN ANIMALS

1. The damage done   53
2. Correcting the balance by biological control   69

PART III

ANIMALS IN THE SERVICE OF MAN

1. Producers of food and clothing   83
2. Beasts of burden   103
3. Companions and assistants   112

PART IV

OTHER ASSOCIATIONS BETWEEN ANIMALS AND MAN

		Page
1.	Animals in sport	127
2.	Animals as providers of goods	138
3.	Hangers-on	153
4.	Animals, man and disease	164
5.	Wild animals in captivity	173
6.	Animals in religion, folklore and superstition	185
	Bibliography	193
	Index	195

LIST OF ILLUSTRATIONS

Ploughing near Ambleside	*Frontispiece*
The passenger pigeon, now extinct, was North America's commonest bird a century ago	*Facing page* 16
The egret's plumes once put it on the danger list	16
The avocet has recently returned to breed in Suffolk	17
The white-tailed gnu, now reduced to a few tiny herds in South Africa	32
In West Africa the bongo has learnt to live near man	32
The birds in this mosaic in a Sussex Roman villa are assumed to be pheasants	33
A grey squirrel robbing a British bird's nest	33
Little owl carrying a field-mouse	36
In Australia the rabbit has become a frightful pest	36
The robin helps to destroy harmful insects	37
The ladybird is a lifelong enemy of green-fly	37
In Ancient Egypt cats were kept to control rats and mice	37
Dense prickly pear in Queensland, Australia	44
The same area three years later after destruction by *Cactoblastis*	44
Milking by electricity	45
Green-fly are the cows of the ant world	45
A family of guinea-pigs in the London Zoo	48
This goat mascot of the Royal Welch Fusiliers was not keen to go overseas	48
There is a big contrast between this wild sow and its offspring and	49
This domestic pig's family	49
Ostriches are still kept on a few South African farms	64
An Egyptian carving, about 2000 B.C., showing a bee sucking honey	65
Worker bees filling the comb	65
A painting of a water buffalo ascribed to the 12th century Chinese painter Liu Sung-Nien	84
Ploughing matches are still popular in many parts of the country	85
A team of pack donkeys in North Ashanti, Gold Coast	92

From time to time zebras are broken in as draught animals	92
Elephants were used in war many centuries ago	93
A relief in the British Museum showing camels used in the 7th century B.C. by the Assyrian king Assur-bani-pal in his North Arabian campaign	100
Peruvian pottery many centuries old shows a bridled llama's head	100
Elephants working at a teak mill in Burma	101
Cavalry on the march: a carving in an Indo-Chinese temple built in the 12th century A.D.	108
Laplanders find the reindeer invaluable for hauling their sledges	108
A team of husky dogs at work in the North-west Territory, Canada	109
A shepherd and his dog bringing in a flock of ewes	109
An Indian ivory carver at work	112
Superb craftsmanship in leather	112
A spaniel being trained as a gun dog	113
The huntsman calling in the hounds after they have lost the fox	113
The Mogul Emperor Akbar hunting blackbuck with cheetah	128
Students at the Field Study Centre of Flatford Mill	129
Feeding London pigeons in Trafalgar Square	140
Brooke's gecko is very much at home in houses in the tropics	141
Two house mice	141
The swallow seldom nests away from buildings	141
The London Zoo in early Victorian times	148
Elephant rides are always popular at the Zoo	148
Feeding sacred ibises in Ancient Egypt	149
Mummies of cats from the XVIIIth Dynasty of Egypt in the British Museum	160
A white horse on a hillside near Thirsk, Yorkshire	160
The arms of Barclay's Bank	161
The Great White Horse, at Ipswich	161
The arms of Martin's Bank	161
The arms of the Royal Veterinary College	161

INTRODUCTION

EVER since man arrived on the earth he has had numerous and varied associations with the animals around him. In the early days these relations were a good deal more direct than they are today in most countries. The Stone Age man killed an animal and then promptly ate its meat and prepared its skin as his clothing; the city dweller of today draws as his ration a piece of meat which may have been frozen for months, his milk has been handled entirely by machinery and comes to him in a sealed bottle, his clothes are made largely of wool grown in Australia or New Zealand, while his shoes, which may perhaps never tread anything rougher than the city pavement, are made of leather. His animal associations by no means end there. If he does not keep a cat or a dog he is more than likely to have a canary, a budgerigar or some other pet about the place; he probably takes an interest in at least one sport featuring animals, be it only greyhound racing, and he quite certainly has need from time to time to take action against a fly or maybe to use a mousetrap. In the course of his life he has probably been protected against smallpox by the use of vaccine prepared from the lymph of a calf. He may well have done some of his military service overseas and there learnt to use a mosquito net in order to avoid malaria and other tropical diseases. Sometimes during his life he has almost certainly visited a Zoo or circus and he is likely to regard a black cat with more favour than a solitary magpie.

But all of these are rather individual associations, in each case a direct relation between a man and an animal. There are also far wider aspects to be considered, where man, as a race, has affected whole sections of the rest of the animal kingdom. In taking effective possession of countries, man has inevitably dispossessed many of the previous occupiers; by bringing in new animals, whether accidentally or otherwise, he has introduced all manner of problems, and his world-wide transport network has helped in the general mixing up. On his wanderings he has taken with him his goods and chattels—

the latter word originally meant what we now know as cattle. But he first had to get his cattle, and the domestication of the larger animals was of fundamental importance in the growth of civilizations—as fundamental, in fact, as the development of arable farming. In countries such as Australia this stock many times outnumbers the human population and the native animals are being driven out by sheer competition.

The more important world-wide aspects of these problems were considered in a series of articles published during 1950 and 1951 in *The Geographical Magazine*. This book is a detailed study of these problems and a number of more individual ones, and it is hoped that most aspects of man's association with animals have been dealt with. I am grateful to Mr. Michael Huxley, Editor of *The Geographical Magazine*, for permission to use some of the material already published and also for his enthusiastic help in suggesting interesting and original illustrations and sources of information.

Until recent years the wild animals had few, if any, advocates, for the immediate needs of man were the only things that mattered. It is good to know that in almost every country there are now both official and private organizations whose main concern is the preservation of our heritage of nature, as, for instance, the Royal Society for the Protection of Birds in Great Britain, the National Audubon Society in North America, and the Fauna Preservation Society, concerned chiefly with the fauna of the British Commonwealth.

Part One

The Results of Competition and Conflict

1. IN GREAT BRITAIN

It has become the fashion to regard man as the lord of creation, the most highly organized animal and one that can impose his will on all lower forms. Unfortunately the usual result is that the lower forms suffer, and for this reason it seems logical to consider first some of the things that have happened when man has found himself in competition or in conflict with the animals around him. When effectively occupying a new country he inevitably has a drastic effect on the vegetation covering it and on the animals living there, even though it may be some time before he makes any impression. As land is taken over for cultivation most of the existing trees and plants just have to go, even though many of them creep back as weeds if given half a chance, and in going they completely change the conditions in which the animals have been living. Some kinds may be displaced because their living space has been entirely destroyed: some may clear out because they object to any interference with their surroundings: in these cases man is in competition with them. On the other hand some kinds may disappear because man hunts them too heavily, whether for food or clothing or because they threaten him or his possessions: here they have come into direct conflict with man. Some animals, however, appear to be more adaptable and they manage to work out a *modus vivendi*; they may even increase greatly in numbers under the new conditions.

In completely settled countries, such as Great Britain and much of western Europe, man's occupation is of such long standing and so complete that a state of equilibrium has almost been reached, and further large changes are not very likely. In new countries, especially in tropical zones, effective occupation is only now taking place and the battle is actively being waged, so that we can watch the changes happening, generally at a pace far more rapid than anything Great Britain ever knew, thanks largely to motor transport, tractors and bulldozers. The North American continent fills a place rather intermediate between Great Britain and the tropics, for though the development was often rapid, most of it took place before these machines were available. In almost every corner of the world the battle has been waged or is still in progress, though both the human objectives and the conditions differ widely: all too often the battle is one-sided, but in many cases a truce has been reached.

Great Britain has now reached a stage in which the only considerable areas untouched by man are those that have not seemed worth touching—the poorest of the moorlands, the rocky hills, the salt marshes—though even here such activities as sheep-grazing and deer-forest management may have materially affected the flora and fauna. As settlement began, the big carnivores were the first to be reduced in numbers; they were too dangerous to man and his stock. Nowadays we can hardly imagine the European brown bear as a native of Great Britain, yet it lived here until about 1,000 years ago. Probably it was not very common—most of the country was not particularly suitable for it—but there are plenty of references to it in the Anglo-Saxon records. In the time of Edward the Confessor, for instance, the town of Norwich was expected to supply the King with one live bear every year, as well as half a dozen dogs to bait it. It is generally assumed that the last wild bears were killed in the 10th century in their strongholds in the Welsh mountains and the old Caledonian Forest in Scotland; man must be held responsible for its disappearance, though he is hardly to be blamed for it.

Several other large animals were lost about the same time. The reindeer had found its range much reduced by changes in climate, but it had managed to hang on in the far north of Scotland, and there is some evidence that bands of hunters used to cross over from Orkney to hunt it in Caithness as late as the 12th century. The elk—generally called moose in America—was also a native of Scotland, and Fraser Darling considers that it had disappeared by the end of the 13th century. Man was again responsible for its going, partly by hunting it and partly by clearing the forests and thus reducing its range. Wild boars managed to survive until much later but they had, of course, been far more common and widespread; in fact, at one time they must have been found over the greater part of the country. As the forests disappeared and the land was brought into cultivation, the wild pigs became rare and latterly were found only in parks and reserves, for they were reckoned almost as royal beasts of the chase. The last Essex specimen is said to have been speared in Elizabethan times by the Earl of Essex, while a few managed to survive long enough in Windsor Park for James I to hunt them in 1617.

Although the beaver quite certainly lived in Great Britain within historical times, there is considerable argument as to when the last one was killed. In the south of England it had probably disappeared before the Norman Conquest, but in Scotland it was known rather later than this, though reports of its being found around Loch Ness in the early 16th century may or may not be true. Place-names derived from the beaver are found in many parts and there is even a Beverly Brook running into the Thames opposite Fulham Palace. It is hardly surprising that we have lost the beaver, for not many parts of the country can have been ideal for it. In western Europe it was once widespread and even common, but it is now found in only a few areas—in Scandinavia, in Russia and in the valleys of the German Elbe and French Rhone: in all these places, however, it is now protected and it is again increasing in numbers.

The wolf was still abundant up to the Middle Ages. The matter-of-fact Baedeker, for instance, records how, in the 12th century, most of the monks of Llanthony Abbey in Monmouthshire, "having 'no mind to sing to the wolves', migrated to Gloucester". In England it was exterminated during the reign of Henry VII: in Scotland it was still a serious menace in the middle of the 16th century and seems to have lingered until the 18th century, while in Ireland a few were reported even towards the end of that century. In *An Account of the Great*

Floods of August, 1829, Sir Thomas Dick Lauder quotes a description by the young Macqueen, Laird of Pollochock, of how he killed the last wolf in Scotland, in the middle of the 18th century:

> "As I came through the slochk (ravine), by east the hill there," said he, as if talking of some everyday occurrence, "I forgathered wi' the beast. My long dog there turned him. I buckled wi' him, and dirkit him, and syne whuttled his craig (cut his throat) and brought awa' his countenance, for fear he might come alive again; for they are very precarious creatures."

There is considerable argument about the status of two other large animals—the wild ox (aurochs) and the wild horse. It may be that they lingered on in Great Britain until historical times as truly wild animals, but the picture is rather confused by their domestication. The disappearance of that beautiful cat the northern lynx was probably the result of climatic changes, and it was almost gone when man arrived in the Highlands; in fact, there is only one authentic record of it—near the hearth of a Sutherlandshire limestone cave.

No other mammals living in Great Britain within the past 2,000 years have been lost, but the status of almost every one has been changed for better or for worse. The red deer, for instance, which is our largest land mammal, has proved fairly adaptable, and in recent years man has taken great care not to let its numbers drop too low; it is too valuable for sporting purposes. In Roman times red deer were found over most of the country and they were largely animals of the woodlands, but as the land was cleared they learnt new habits and they now survive in numbers only on the Scottish moorlands almost devoid of trees. A recent estimate for the whole of Scotland is no less than 150,000 head. In early days, too, man regarded the red deer as valuable prey and hunted it for venison, skin and antlers. Down in the south they managed to go on making a living in parks and forests. Truly wild deer lived in Epping Forest until 1827, when the last few were taken to Windsor Park. Wild red deer are still found in the Lake District and in parts of Devon and Somerset, with a very few in the New Forest. Those living on enclosed parks stem

From J. J. Audubon's "The Birds of America" (British Museum)

The passenger pigeon, now extinct, was North America's commonest bird a century ago

The egret's plumes once put it on the danger list

Paul Popper

Ian M. Thomson

The avocet has recently returned to breed in Suffolk

in part from the native red deer but most of them owe their spectacular size and antlers to the introduction of other races, especially from the Continent, and the antlers particularly are benefited by special feeding.

At the other end of the scale there is the red squirrel, which also has had its ups and downs, but these are not entirely due to human activities. It is probably most at home in the pine forests, and the deforestation of the Highlands brought red squirrels there to a very low level: by the end of the 18th century they had in fact disappeared from many areas. By this time, however, the larger landowners had started to plant new forests and the squirrels began to pick up, probably aided by introductions of Continental stock. Now the wheel has turned full circle; the red squirrel in Scotland is common enough to be considered a pest in the pine plantations and it is likely to hold its own in spite of control by squirrel clubs.

In the south the native squirrel got along fairly well in the mixed woodlands and it must have been abundant in the early part of the 19th century, for it is recorded that most of the 20,000 or so sold each year for eating had been taken around London. This heavy toll did not check it, for it continued to increase until about the beginning of the present century, when an epidemic almost exterminated it. At this stage the grey squirrel came on the scene and was able to take over much territory with virtually no opposition; the grey squirrel is without doubt a frightful pest, but it is probably unfair to blame it too much for the disappearance of its red cousin from many parts. More recently the red squirrel has recovered some ground in England and it is certainly in no danger of extermination.

The pine marten and the polecat are two carnivores that have come through many vicissitudes and, like the red squirrel, capercailzie, crossbill and the crested tit, are now likely to find the extensive new moorland and highland plantations much to their liking. The marten suffered first a great reduction of range as the forests were cleared and then the gamekeepers stepped in; W. H. Pearsall quotes trapping figures in Glen Garry for a three-year period 110 years ago and the list included not less than 246 martens and 106 polecats. The marten, however, then changed its manner of life, somewhat as the red deer has done, and took to the more open ground, especially in north-west Sutherland; it chose this area more wisely than it knew, for the landowners proved sympathetic and organized its protection. From there it has spread to several adjoining counties and, with direct human assistance, has established itself again in the southern Highlands. Given reasonable protection it should now be safe.

The polecat on the other hand failed to survive in Scotland: perhaps this was because it travels less and it is probably easier to trap on the ground. In the Welsh hills and mountains, however, it is fairly secure and it has always been much more plentiful there than the marten. Colin Matheson has collected records of no less than 500 Welsh polecats killed in the five years before 1932; it has probably increased considerably since then, due in part to the lack of gamekeepers during the war and also to the increase in forest area, so that this also is in no danger at present.

When we turn to birds we find a rather similar story but with one essential difference—the birds' powers of flight always give them a chance of returning to our island if conditions should change. Even without direct human aid in restoring it the loss of a species may therefore be only temporary.

In mediaeval days the fens and other marshlands were dreary wastes, providing ideal hiding-places for such figures as Hereward the Wake, the breeding grounds for innumerable marsh birds—the bittern, avocet, black-tailed godwit, marsh and Montague's harriers. The 18th and 19th centuries saw the conversion of these marshlands into some of the most fertile farming areas in the whole country, with such curtailment of potential breeding grounds that, with the assistance of gamekeepers and collectors, all of these beautiful and interesting species had virtually been lost as breeding birds by quite early in the present century. Fortunately the story does not end there; the past fifty years have seen a big change in our attitude towards our heritage of nature, and the protection granted has been the chief factor in bringing about the return of the bittern, the avocet and the harriers as regular breeding species, though most of them nest only in very small numbers, while several others appear to be taking an interest in possible breeding places.

The story of the osprey is rather similar to that of the marsh birds, for it disappeared as a breeding bird early in this century, though it traverses Great Britain regularly as a passage migrant. The osprey is a big fish-eating hawk which catches its prey by plunging into the water; it is thus conspicuous and vulnerable and its loss was very largely due to persecution by gamekeepers and greedy collectors. But it is a bird whose niche in nature is a narrow one and even with rigorous protection it might well be unable to re-establish itself. A very different factor was the chief cause of the kite's disappearance. In the course of time this bird of prey has become largely a scavenger and in one form or another, all quite similar to the British variety, it is a common sight in the neighbourhood of towns and villages of all parts of the tropics. In the days when the usual method of garbage disposal was to dump it in the street the kite played a valuable part in sanitation, and up to the 17th century it was very common in and around London. It was still common in many parts of the country nearly a century later and Gilbert White recorded seeing several kites on the Sussex Downs in 1773. In *The Winter's Tale* Autolycus gave warning "when the kite builds look to lesser linen", which suggests something of the town habits of the kite in Shakespeare's day. The advent of the dust-cart and more hygienic methods of rubbish treatment by burning, burying and so on, squeezed out the kite from the position it had found; gamekeepers, working on the theory that any large bird of prey naturally kills game, reduced the kite's range to a few inaccessible parts of Wales, where egg collectors then competed to take the last British kite's egg. A tiny handful of kites still survives, thanks to the most strenuous and unremitting protection of individual nests during the nesting season. It is doubtful whether the kites will finally be saved, and one rather wonders whether such expenditure in money and manpower is really justified.

Most of the larger birds of prey have inevitably been reduced in range and numbers, though the only breeding species actually lost—apart from the osprey—is the sea eagle, which last bred in the Shetlands in 1908: the male of this pair was then shot and the female, unable to find another mate, hung around the nesting site for several years. The sea eagle was never very common but it had bred in the Lake District and in the Isle of Wight until towards the end of the 18th century.

The development of organized gamekeeping was the biggest cause of this reduction in the big birds of prey and the numbers killed in parts of Scotland around the beginning of the 19th century are almost beyond belief. W. H. Pearsall gives many details in *Mountains and Moorlands*: one report quoted refers to the Duchess of Sutherland's estate from 1831 to 1834, when 224 eagles and 1,155 hawks and kites were killed. These perhaps seem fantastic figures for a period of only four years and one cannot help wondering what they all fed on. But the kill of a number of estates at that time were equally staggering and the countryside must have looked more like parts of the tropics today, where the sky is never without its large birds of prey. Nor was that the whole story, for on the same estate in the same period they also list some 900 ravens, 200 foxes, 900 wild cats, polecats and martens, and 1,700 crows and magpies. For several species of both mammals and birds the kill in those four years would probably equal the present-day population for the whole of Great Britain! It should perhaps be mentioned that some authorities today consider that these figures may have been grossly inflated because keepers depended largely on the bounties for their living: even so there must have been many birds and beasts of prey in those days.

To the gamekeeper most birds of prey except owls and kestrels are usually vermin, but the problem is not quite so simple as that, for some of these birds are deadly enemies of rats and mice, both of them unmitigated pests. The equation is indeed a complicated one, with so many unknown factors that even the experts differ after years of scientific investigation. This is especially true of the birds of prey but there is little argument about two vermin of the crow family—the magpie and the carrion crow. Farming activities combined with game preservation offer these two birds fine opportunities of making an easy living and throughout the spring and summer they are serious enemies of birds' eggs and young. Between the wars they were kept well in hand by the gamekeepers, but during the Second World War practically all control was taken off.

Crows and magpies can seldom have been so numerous as they then became and six years after the war ended there was still no sign of any reduction in most districts. The name "carrion crow" is no longer apt, for it takes live food when it can get it and does terrible damage to the small—and not so small—birds of our countryside. The buzzard has also become much more plentiful and will probably stay so, for on balance it does much more good than harm.

Perhaps it is appropriate to mention the serious egg collector next to the crow and magpie, for he combines their destructive and thieving habits. Drawers full of eggs may be necessary in a Natural History Museum but they can give the average person little pride of possession: the real snag is that the rarer the bird becomes the more its eggs are wanted, and the egg collector is then added to those enemies that have put it on the wanted list. Egg-collectors are getting rarer themselves, which is all to the good, but it is an appalling indictment that even now the nests of some extremely rare birds have to be protected individually against their wholly illegal efforts. Small boys do some damage in the "bird's-nesting" season, but such damage is indiscriminate and unlikely to endanger a species, even though some birds on the outskirts of towns may find difficulty in rearing families.

The capercailzie is an example of a bird whose tremendous fluctuations have been entirely caused by man. Although not limited to actual pine trees, it does depend on conifers of some sort for much of its winter feeding. Drastic cutting of the old forests reduced its range, and its large size—for it is nearly as big as a turkey—made it easy to hunt, with the result that the last native bird was killed in 1770. Several efforts, the first unsuccessful, were made to reintroduce Swedish birds, and those put down by Lord Breadalbane at Taymouth Castle in 1837 were the ancestors of the present large stock. By that time woodlands were being created again and the caper soon spread over several counties. The mass planting projects of a century later will in time give it still more scope but now it is more or less on the vermin list of the Forestry Commission! But the areas concerned are large and it should be safe enough, even though regularly shot.

The populations of most animals vary continually, though

some show much greater fluctuations than others, but it is quite unfair and misleading to blame man for all the cases where numbers have fallen to a seriously low level. The climate is sometimes responsible: several severe winters in recent years, for instance, were very great setbacks to the beautiful little bearded tits. Quite often, however, no reason can be suggested for the change: the wryneck is now a very rare bird—though I can remember it as really common in some parts of Essex in the 1920s—but nobody knows why. Nor do we know just why the curlew and redshank have greatly increased their breeding range, and the black redstart begun to nest in England. We are perhaps tempted to blame mechanized farming and artificial manures for the reduction in numbers of some animals: the use of the self-binder and harvester has made the harvest mouse very rare, while frequent mowing cannot but disturb the breeding activities of lapwings and partridges. This is especially true in areas where the thick hedges have been grubbed up, so that gamebirds are forced to nest out in the crops. At the same time modern agriculture does not seem to be responsible for the near disappearance of the corncrake, likely as this may seem at first sight.

The famous great auk was lost to both Great Britain and the world through senseless slaughter. Found only in the northern Atlantic, the great auk was far and away the biggest of the auk family, of which several kinds are so typical of many northern rocky coasts. It could not fly at all, yet it migrated south in winter, sometimes reaching the Bay of Biscay: it nested on several rocky islands near Iceland and on Funk Island near Newfoundland, where, when it was discovered in 1534, it existed in thousands. Slaughter went on with little intermission and the last was killed about the middle of the 18th century. It seems that these unfortunate birds were killed only for their feathers, the carcasses being left to rot. Today the great auk, or garefowl as it was commonly called, exists only as a few skins in museums, and also a few score eggs—the only bird on the British list in historic times to become extinct.

In the seas around our coasts we can see, in over-fishing, the result of misusing natural resources. A given area of sea can produce only a certain maximum amount of fish (apart from the markedly migratory kinds such as herring), just as one area of land has its productive limits. If more is caught each year than has been produced, the stocks fall and the average size also falls—finally to a point where fishing is no longer profitable. Stocks of fish in the main fishing areas around the British coasts fell steadily between the wars, and more and more vessels were compelled to go to distant fishing grounds. The war brought respite to many sea areas, especially in the North Sea, and the quantity of fish taken and their average size went up a lot when these grounds were again opened for fishing. The most practical control on our fishing is probably the fixing of a minimum net size, through which most of the younger fish will pass. The lobster industry on the Atlantic coast of North America is an interesting case of successful control by size. When lobsters became popular there was an immediate increase of fishermen, but people soon realized that the lobster might almost be wiped out and so a minimum size-limit of nine inches was imposed. This was enforced for some time, but it did not help the lobsters much, for they had not really started spawning at that size. So the limit was increased to $10\frac{1}{2}$ inches and the lobsters at once benefited; so did the fishermen, for anything up to 17,000,000 lobsters a year are now caught on the Maine Coast.

The real snag in these matters is to get international agreement about action in non-territorial waters: everybody seems to wait for the others to take the first restrictive step. The North Sea area is particularly difficult because of the variety of countries interested, but the U.S.A. and Canada have shown what can be done in the common good. Some years ago the halibut fisheries in the north-east Pacific were obviously being hit too hard: in 1930 both countries agreed on a quota system and fishing stopped for the year when the agreed weight had been taken. The halibut responded and within a few years it was reported that the quota was being filled in five months as against nine months when the agreement was made. The fishery is again profitable and it is even more possible that a slightly higher quota can be fixed.

The sea fish off our coasts were heavily reduced by direct attack by man; he took a heavier toll than the population could support. But in many of the rivers a heavy loss of life and productivity has resulted from a by-product of civilization—river pollution. There are many forms of this evil, but the two most important are the discharge of waste products from industrial premises—whether actual chemicals, dirt or just hot water—and the effluent from sewage works. In some cases the fish are killed directly: in others the fish and the minute forms of life on which they feed are slowly poisoned or starved of oxygen: but the result is the same—the river becomes unproductive. It is perhaps hard to imagine anglers on the Thames at London Bridge today, but early last century salmon were still running up from the sea. The first pollution was caused by the escape of all sorts of chemical waste from the newly constructed gasworks and the discharge of raw sewage, one of the things most calculated to use up oxygen and make things difficult for the fish. Since then London has become the biggest port in the world and this involves a great volume of assorted pollution which can hardly be avoided, but it is good to note that the past few decades have brought a steady improvement, especially in the way of refuse disposal. Solid matter is either burned, dumped in special tipping areas or taken right out to sea; some authorities convert it into heat and innocuous ash at little or no cost. Similarly, in modern methods of liquid sewage treatment the outflowing water is rendered harmless to animal life and, in being broken down, the organic matter produces more than enough gas to run the station's power plant. As a result, animal life in the Thames is picking up again: sea fish and other creatures penetrate farther up the tidal water and freshwater varieties farther downstream than they did fifty years ago. Gulls and ducks are also finding the cleaner river a greater attraction.

We cannot end this discussion of man and his effect on British animals without noting one or two ways where he has helped rather than hindered. Sewage farms have proved welcome ports of call for passage migrants, especially the wading birds and the lovely black terns: the water-supply reservoirs, especially in the vicinity of large cities, have become a great attraction for many ducks and serve as real sanctuaries. Other large reservoirs, formed in the mountainous areas in connection with water supply or hydro-electric power schemes, have had a big bearing on fish resources. Most of these new waters are very suitable for trout and salmon, and some of them have proved astonishingly productive. On the other hand the formation of some reservoirs seriously interfered with the run of salmon going to the head waters for spawning, but it has generally been possible to provide ladders by which they could by-pass the falls.

We are most used to seeing the charming family of wading birds in the open spaces, whether it be in the autumn on the saltings and sea-shore, or in the breeding season on the moorlands, marshes and shingle banks. Quite a number of waders find the sewage farms convenient stopping-places on their migrations but, in general, they are an independent lot as regards man. There seems to be one exception in the person of the little ringed plover, the latest addition to the list of regular British breeding birds. The first nests had been recorded in the shingle banks around some Middlesex reservoirs, and an intensive survey in the years just before the war revealed other nests in similar spots. When a big new reservoir was being constructed in the Chingford area its rough floor—the working area for numerous tractors and bulldozers—provided a happy hunting and nesting ground for several pairs of these delightful and spruce little birds. They seem, in fact, to specialize in just such places both here and in western Europe, from which our colonists have no doubt come; it is recorded that extensive canal digging, with resultant large stretches of gravel, attracted them to parts of Belgium where they had never been seen before, and they almost seem to depend on man's activities for their breeding grounds.

The gannet's increase is explained in a very different way—by a change in human feeding habits! Gannets were formerly eaten in large numbers: in some of their breeding colonies the "harvesting" just kept level, but in others it steadily reduced them. When cheap imported meat became available the demand for gannets dropped: further, some islands were evacuated and the pressure taken off completely—in the case of St. Kilda the annual toll had been as high as 22,500 birds. The gannet quickly responded and the British population rose from some 70,000 in 1889 to about 110,000 today. But imported meat is no longer cheap: perhaps we may have to start eating gannets again!

2. IN AFRICA

AFRICA is a vast continent in which one finds every type of country, from desert waste to the most luxuriant rain forest and the most delightful temperate zone woodland. Two thousand years ago North Africa was comparatively civilized, though most traces of that civilization were obliterated after a very few hundred years, but most of the continent, apart from the narrowest coastal strips, remained virtually untouched until less than a century ago. It is true that many and various African tribes were making a living there, but they lived fairly self-contained lives, at equilibrium with the surrounding bush and its animal inhabitants, until the coming of the European.

For many parts of Africa the past century—even the last twenty years perhaps—has brought a change from untouched bush to fully mechanized farmland. Man has inevitably had a tremendous effect on the animals, and we can study here the sort of thing that was going on a thousand years ago in our homeland—but speeded up beyond measure—and we too often have to look on at a process we have in fact started but yet are powerless to direct or stop.

There is little uniformity in Africa; the conditions in East, South and West Africa and the problems they pose are quite distinct, and within each of these rough divisions (each bigger than western Europe) there is infinite variety in topography and climate and, therefore, in vegetation. Development, too, is very irregular, and there are large areas where man has so far failed to have any effect. Areas too inaccessible or dry or infertile to farm profitably, freshwater swamps, desert or mountain regions, open woodlands where the tsetse fly is still a tyrant forbidding humans and cattle to thrive: the occasional hunter may penetrate to such places, but he has no more effect than any other natural predator, such as lion or leopard, that goes there to hunt. But throughout the great Dark Continent the frontier of civilization (if we can call it that) is still moving, and the areas of untouched country grow less and less. First the inaccessible fertile areas are opened up; then the attack shifts—to irrigate the arid zones, to dispossess the tsetse, to fertilize the infertile. Every move affects the animals and, for the most part, it affects them adversely, even disastrously, though there are notable exceptions.

The early explorers spoke of the plains alive with big game of a dozen varieties—the plains and plateaux so typical of much of East, Central and South Africa: what has been happening to these hosts of large animals as they came in competition and conflict with man, when his frontier was no longer static? It is the fashion to blame the white man for most things that happen in colonial territories, but it was actually his good deeds that indirectly caused the first pressing back of the wild life in some areas. One of the factors most responsible for keeping the native population at a low level was slave-raiding: this was stopped by European action and the immediate result was a much more settled existence and a growth in population, demanding more food, which, in turn, called for larger cleared areas. Medical and public health services, especially in the British territories, coupled with the spread of Christianity and the wiping out of some crippling superstitions, have steadily lowered the African death-rate, especially the infant mortality, which had been appalling. This resulted in an even more marked rise in population, with progressive occupation of new land, if available, or more intensive use of the old.

At the same time the African has found an appetite for imported goods such as clothes, sewing-machines, cigarettes, sugar and paraffin, and he has to grow crops in order to get his purchasing power: these are known generally as cash-crops —such as maize, tobacco and pyrethrum in the open country, and cocoa, palm products and bananas in the forest zone. So the pressure on the land builds up. In many areas, however, there has also been the development of European farming on a fairly large scale: both directly and indirectly this adds greatly to the pressure. In the more or less temperate zone of South Africa colonization by the Dutch had started at the beginning of the 17th century, but it was not until at least two centuries later that the colonists made themselves felt: for the most part, even then, they were intent on subsistence farming and the days of mass food production for export had not arrived. Serious European settlement in the East African highlands did not start until early in the present century. But the battle has been joined more or less everywhere, for large herds of game and extensive farming cannot exist in the same area: it is a case of direct competition for the use of land and the animals are bound to lose.

The first total loss among the animals had occurred in North Africa—a wild ass known to the Romans and featuring in several mosaics and pictures. Apparently it was confined to the neighbourhood of the Atlas Mountains and was still in existence about A.D. 300, but a little later it disappeared, primarily because it was hunted by the peoples who had colonized the southern shores of the Mediterranean.

There was a long interval before the next casualty had to be recorded, and this was at the other end of the continent. What is now Cape Province was the first area to be effectively occupied by the European settlers and they found the blauwbok, or blue-buck, living in a very small corner of the Province: some observer suggested that their natural *lebensraum* was only a very few hundred square miles. At first it was so common within this restricted range that it was well known as the "blue goat", though this was not a very apt name for a smaller cousin of the very handsome roan antelope. By the second half of the 18th century it had become rare and in 1800 the last few blauwbok were shot. Only five museum specimens remain today, none of them in Great Britain. Man was directly to blame for this loss, but its extremely limited distribution had made it very vulnerable.

The quagga came next and this has had more publicity than any other animal that has vanished. It was a zebra living on the open plains of Cape Colony and the adjoining parts of Griqualand West and the Orange Free State, and at one time it was found in huge herds. The name "quagga" comes from the Hottentot and was intended to represent its barking call. Scientists recognized the quagga as the partly striped form, half zebra, half donkey, but in the vernacular this name was, and is, applied to zebras in general, and this is the explanation of the periodical reports that the quagga is not extinct.

The quagga was first described in 1758 and it was still common enough fifty years later, but by that time the drive had started and the Boer farmers must be held primarily responsible: they wanted the land for their own stock, but they also attacked the quagga directly, feeding the meat to their farm labourers and shipping out the hides literally by the wagonload. By the middle of the 19th century the quagga had virtually disappeared from Cape Colony. The last wild quagga is said to have died about 1878, but the very last of the species died in the Amsterdam Zoo in 1883. The London Zoo had three specimens, one of which lived until 1872, and this seems to have been the only live quagga ever to be photographed. Had sanctuary been granted it within its natural range the quagga would never have been lost, but a century ago the need for wild life conservation had not yet been appreciated.

Meanwhile several of the big carnivores had been seriously affected by the disappearance of the herds of game, as well as being attacked directly. The Cape lion was lost about 1865 and the Barbary lion in the early 1920s, but they could be regarded as local forms of the African lion, which itself is in no danger at present.

The next large animal to go was a race of Burchell's zebra, the southernmost form of the zebra which is found right up into Kenya. (There are two other quite distinct types: the mountain zebra in the south and the Grevy's zebra in the north-east.) Its history was very like the quagga's: at one time immensely numerous, though confined to the Orange Free State and parts of Bechuanaland, it suffered at the hands of the colonists, especially the hide-hunters, and was virtually extinct by the end of the 19th century. Three other forms of this zebra are still common enough to be a nuisance in some parts of East Africa and this loss is much less serious than the quagga's. There was a time in South Africa when both the quagga and the southern Burchell's zebra were seriously considered for domestic use in both South Africa and Great Britain. As Menzies points out, had not the invention of the petrol-engine revolutionized road transport, these two zebras might well have been thought valuable enough to save for transport purposes.

In addition to those gone beyond recall there are several animals which are no longer found in the wild state but exist today thanks chiefly to vigorous protection on various estates in South Africa. The bontebok was always restricted to a small part of Cape Province, while the blesbok, another closely related hartebeest, ranged through the Orange River Colony, the Transvaal and Bechuanaland, at one time in very great numbers. The former was saved by the action of the Van der Byl family and is now preserved in a special National Park. The blesbok, formerly one of the most numerous antelopes in Africa, had been nearly exterminated by the end of the 19th century; the only herds in existence were on Dutch farms and during the Boer War these were almost wiped out. Now, however, a very different factor has intervened to save the blesbok, for it has become the chief source of venison in South Africa and is protected on farms as a meat-producer: such a project may be very little different from cattle-ranching in America and Australia but the net result is that the blesbok is protected.

The white-tailed gnu, or black wildebeest, is one of the most grotesque animals that ever existed: it is, in fact, a caricature in build and movements, and its noises are equally fantastic—especially a snort which is almost a bark. Like the blesbok it was once very plentiful, but it had been reduced to a few farm herds: the Boer War helped rather than hindered, by releasing and mixing herds, but they were all subsequently enclosed again. This gnu, however, does not seem to have the same capacity to survive as the blesbok, and in spite of Government protection its future is by no means certain. There are small groups in several European Zoos, including Whipsnade, but they maintain their numbers with difficulty.

On the plains of Central and East Africa things have not quite got to this pass, and no species seems on the verge of extermination, but everywhere the total numbers of big game are steadily decreasing: this decrease is inevitable, the natural result of curtailment of range as man takes over more and more land for his needs. In addition, within this reduced range the game is under more pressure from native trappers and hunters as well as European sportsmen, though in fairness it must be agreed that the true sportsman, as against the meat-hunter, does little if any serious damage. Even so, there seems to be no immediate danger of any species being exterminated, owing largely to the Reserves and National Parks which are sufficiently well distributed throughout tropical Africa to take care of most kinds of game. The day will come when the only large herds of big game will be found within these Reserves and parks; it will not be just yet, for, as is all too evident in the areas being developed in East Africa for groundnut and sunflower-seed production, some combinations of climate, soil and vegetation strenuously resist effective occupation by man.

When we turn to the closed forest zone of West and Central Africa we find conditions completely different from those in the more or less open country, and the pattern of change, too, is very different. This forest, stretching almost unbroken from Sierra Leone in the west through to Nigeria and Gabon, then across the Congo to Uganda, varies infinitely in composition, but it is always complex, with many different kinds of tree, shrub and herb to the acre. It is called "closed forest" because its various layers, either individually or jointly, form a closed canopy and the floor of the forest never has the dense carpet of tall grass, burnt annually, that is so typical of the open plains. The tallest trees reach a height of perhaps 150 feet. This forest zone generally has an annual rainfall of between 50 and 80 inches, often with two well-marked peaks, and a comparatively short dry season. A few areas, especially in Sierra Leone and the Cameroons, have a rainfall of up to 200 or 300 inches per annum, but this is rather exceptional.

The problem of forest destruction is more or less the same in all parts of this great block, as well as in forest patches in other parts of tropical Africa, but the following remarks refer particularly to the Gold Coast forest, where I spent some fourteen years as a Forest Officer and took especial interest in the results of human interference on the forest—and its inhabitants. By comparison with the African plains and Indian jungles this forest seems very inhospitable: in many areas food plants are almost non-existent and at most seasons there is virtually nothing for man to eat—either roots, leaves or fruit. A few centuries ago the West African forests were practically uninhabited, certainly the few forest tribes made no real impression on them, and it was only the introduction of a series of staple food crops—Indian corn, cassava, plantain, yams, citrus fruits and many others, mostly from the West Indies—that made possible the effective occupation of the forest. To animals other than man these forests are also rather barren feeding grounds. The dense undergrowth makes them unsuitable homes for the large horned antelopes, and even the dwarf antelopes or duikers, the typical small ruminants of the forest zone, are never very common, neither are the forest pigs. The complete absence of any carnivore corresponding to the tiger and the general scarcity of the leopard in this zone are additional facts suggesting the poverty of the African closed forest.

When man does occupy the forest and farm it he changes things far more drastically than in the open country. Dense, perhaps impenetrable, forest is replaced by large open clearings with a few scattered trees, and the whole structure is altered for years, if not permanently. Yet man's power over nature is limited and he has yet to work out a satisfactory system of permanent food-farming in the West African forest: when an area has been cropped for about three years it has to be abandoned and the forest is allowed to grow up and take control once again. The much-famed fertility of tropical forest is often purely superficial, the result of taking out in a few years' crops the richness that nature has built up slowly. A rest of several years, with the re-formation of forest, is then needed to restore some fertility: and this is a most important factor for the fauna, for this regrowth, or secondary forest, is much more favourable for most kinds of animal life than virgin forest. Remains of farm crops provide roots, herbage and perhaps fruits eagerly sought for by many animals, and the population rises rapidly. The oil-palm, whose fruit is a favourite food for innumerable birds and animals, becomes established, together with several other colonizing trees with edible fruits; the popularity of their fruit is the reason for their quick occupation of abandoned farms, for the seeds go everywhere in the birds' and bats' droppings. Such areas are normally cleared again for farming after ten years or so, by which time the soil has

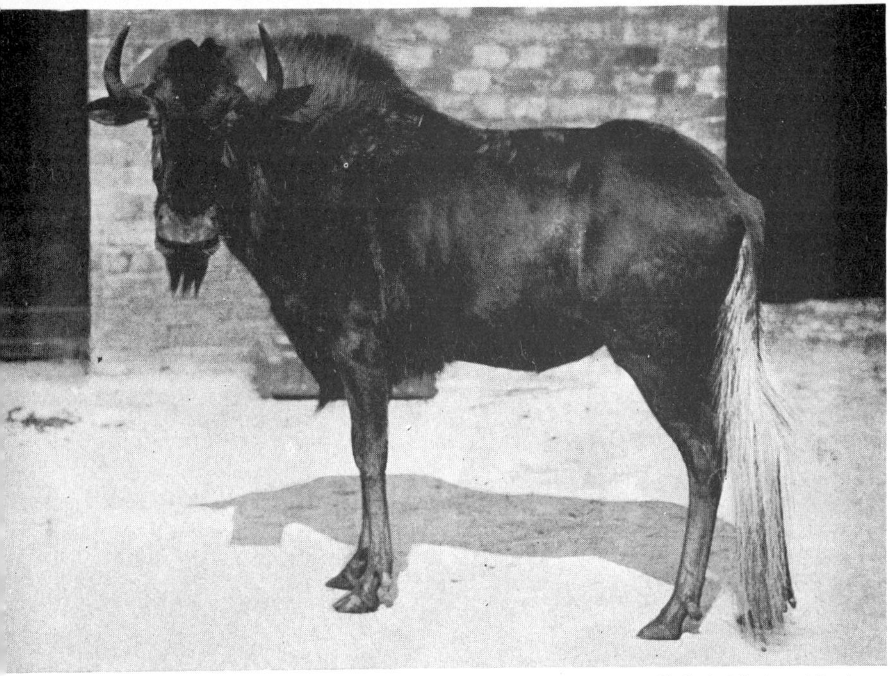

Zoological Society of London

The white-tailed gnu, now reduced to a few tiny herds in South Africa

In West Africa the bongo has learnt to live near man

Paul Popper

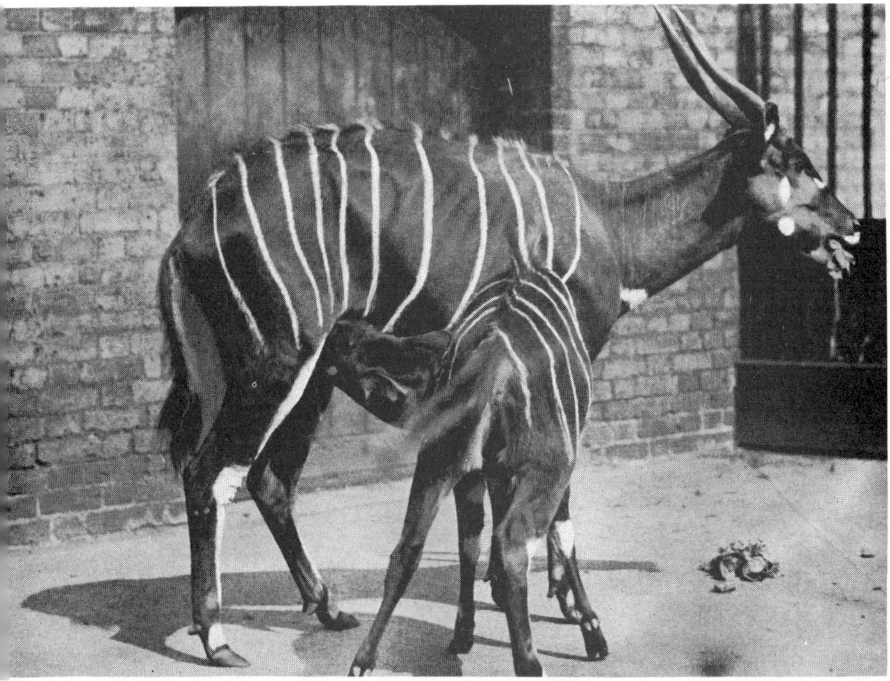

Walter Gardiner

The birds in this mosaic in a Sussex Roman villa are assumed to be pheasants

A grey squirrel robbing a British bird's nest

Walter J. C. Murray

once more become enriched, but there is, in effect, a long-term rotation, so that new areas of regrowth are always becoming available and the animal colonists need never be homeless.

The first clearing drives away some of the animal life, especially such species as the leaf-eating Colobus monkeys whose home is in the treetops of the untouched forest: but the secondary forest zone has a denser animal population, including many small kinds that seem hardly able to make a living anywhere else. This applies particularly to the squirrels, of which the Gold Coast forest has no less than eight distinct kinds. Five of these are seldom if ever seen other than in farm or secondary forest areas and a sixth, though living up in the treetops, is fond of various farm crops and often comes into cultivated areas. Of these five squirrels two make their homes mostly in the medium-sized trees left scattered in the farms for shade purposes and two others keep to the tangled undergrowth: they take a variety of food but their basic diet seems to be the fruit of the oil-palm. The last member of the quintet is the ground squirrel, and it is almost certainly a recent arrival from the open country farther north. As man made the clearings, especially large clearings, connected by tracks or roads lined with farms, so the ground squirrel came in and occupied them, for it found the farms much to its taste. But it does not seem able to penetrate the inhospitable forest and thus reach the isolated clearings, however large and otherwise suitable.

Many other rodents besides the squirrels flock into these old farmlands, rodents with such expressive names as the rusty-nosed rat and the three-striped mouse. Even the forest-loving black-and-white flying squirrel (a handsome creature measuring nearly three feet in total length) has discovered that one particular tree (*Terminalia superba*), known in the timber trade as Black Afara, attains just the right sort of hollowness to serve as a roost after being fire-damaged in farm-clearing operations. One almost expects this of the rodents, for they are an enterprising tribe and include those most versatile pests the rats, mice and rabbits, but it is much more surprising to find some of the larger animals taking advantage of man's activities and more or less carving out a new niche for themselves. But this is what the bongo has done in the Gold Coast. Known in East Africa as an inhabitant of the mountain forest, it is rightly considered one of the shyest big-game animals and one of the most handsome trophies in the whole of Africa. Comparatively little is recorded about it, but observers in Kenya seem to agree that the bongo there is intolerant of any interference. In the Gold Coast its temperament seems to be very different, though nobody would suggest it is not shy: there it has come to find a new home in the abandoned farmlands, with their remnants of food plants such as sweet potato, cassava and cocoyam, and the dense thickets. It feeds by night in farms and old farmlands and hides up in impenetrable undergrowth from well before dawn until after dusk, even surviving in some quite heavily populated areas after most other animals, big and small, have been killed and eaten by the farmers, every one of whom has a gun.

In some parts of West Africa the growing of cash-crops such as oil-palm and cocoa—both of them permanent tree crops—has greatly increased the speed at which the total forest area is decreasing. In this forest zone, too, the human population is continually growing and, even though this increase largely settles in the towns, the result is a greater demand for food and thus a greater area comes under cultivation. The day will come—it has come already in some areas —when the only remaining forest worthy of the name is in Forest Reserves, but as it is not yet possible to keep land permanently cleared for food farming, the patches of fallow will still provide plenty of cover for the smaller forms of wild life: on the other hand, meat is often a rare luxury in many parts of the forest and even mice and bats are hunted for the pot, so that man may finally destroy all the animals that he started by helping unconsciously. In most forest areas there is an adequate framework of Forest Reserves which are likely to remain intact unless the political situation deteriorates and in time, of course, all of these Reserves will be worked regularly for timber and other produce. It has been thoroughly proved in many countries, especially in India, that game can live more or less undisturbed in properly worked forests, and if enough well-distributed Reserves are also declared breeding sanctuaries the medium-sized and larger animals of the West African forest should be safe enough. The allocation and maintenance of these sanctuaries would certainly be a very unpopular measure at first, but such action is quite practicable. The West African people understand the principle of conserving natural resources and apply it themselves to such activities as snail-gathering: they are quite capable of applying it to larger deer. Many fascinating small creatures are found in the African forest—I worked among them and studied them for years, so it is only natural that I have a particular affection for them—and it would be a great tragedy if they just went the way of the dodo, the passenger pigeon and the blauwbok.

In some parts of Africa man has unwittingly upset the balance of nature by killing the big carnivores: in parts of East Africa, for instance, leopard skins became in such demand, especially during the war when the numbers of Europeans and Americans in many territories were greatly increased, that leopards and servals were almost exterminated locally. In open country the leopards prey largely on baboons and are possibly their most important control: with the leopards killed off the baboons have now become such serious pests in some cultivated areas that drastic measures have to be taken against them.

In parts of Sierra Leone several kinds of monkey have got right out of hand in recent years: there is little doubt that the tremendous increase is largely the result of man upsetting the balance, but several different factors come into it. Farming operations have completely changed the vegetation over large areas: the secondary bush seems more favourable to all monkeys, except the treetop leaf-eating group, than the virgin forest and it provides ideal cover from which to raid the food and cocoa farms with which it is honeycombed. If this great increase in monkeys occurred in the Gold Coast forest there would be no complaints, for monkey is a popular meat and man would keep them down to a reasonable level: in Sierra Leone, however, there seems to have been a shift in religion from the previous fetish- and ancestor-worship to a rather vague Mohammedanism, which forbids the eating of monkey. Man is no longer a predator, and drives have had to be organized to save the farms from being completely wrecked, and upwards of 10,000 monkeys may be killed annually. It is very unusual for monkeys to multiply like this and Sierra Leone seems to be unique.

3. IN NORTH AMERICA

THE North American continent contains every type of country from tropical desert to Arctic waste. There is a wide area not very unlike the temperate zone woodlands of western Europe, and within this area it is possible to see a speeded-up version of the sort of thing that occurred in this country as it became effectively colonized. North America provides one or two spectacular stories of man's conflict with nature and we find in the American bison the most striking example of a large animal being brought to the verge of disappearance by human action and then saved by the same agency. Although is it still commonly referred to as "buffalo", the bison is a very different beast from the true buffaloes found in the tropics of the Old World: these all have short smooth hair and they are typically swamp or riverside animals.

First accounts of the bison were brought back from Mexico by Spanish explorers. In early colonial days it was found from New York State down to Georgia and right across the western plains; and on their annual migrations the bison of the plains reached as far north as Lake Winnipeg. Huge herds of another very closely related race were found in the forest regions of Canada, even as far north as the Great Slave Lake. The herds of North America are said to have totalled, at one time, more than 60,000,000 beasts. It is hard to say just when numbers began to fall, but by about 1800 the bison had been cleared from the eastern States. There were two main reasons for this war on the herds: they grazed on areas required for farming

Little owl carrying a field-mouse

Eric Hosking

In Australia the rabbit has become a frightful pest

Camera Press

Ronald Thompson

The robin helps to destroy harmful insects

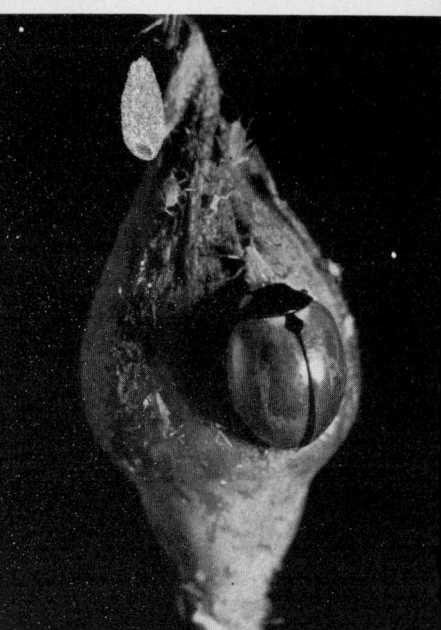

Fred H. Wylie

The ladybird is a lifelong enemy of green-fly

Sidgwick & Jackson

In Ancient Egypt cats were kept to control rats and mice

or stock-grazing and they were a source of meat and hides. In the middle of the 19th century about 250,000 were still being killed every year in the Missouri region alone, and even in 1878 a large drive resulted in 100,000 hides; but by the late 'eighties the bison was approaching extermination, and at the end of the century there were only two herds of wild bison in the whole of North America, numbering barely 1,000 animals. Active protection was started at the beginning of the 20th century and by 1942 the U.S.A. had over 4,000 head, scattered over forty-one States, while a much larger stock had been built up in Canada. Today, with numbers up to 40,000, the American bison is safe enough: it has indeed reached the stage where herds have to be kept at more or less their present size by regular culling for the market. The saving of the American bison has been a wonderful achievement, but it would hardly have been possible without the wide, rather empty spaces still available in that great land. The story of the European bison, or wisent, is told in another chapter.

The beavers are rather like the bison in having two branches of the family, one in Europe that has fallen on very evil times and one in America that at one time got low but was saved in time and became completely re-established. The beaver never came anything like so near the danger line as the bison, but their relative position in the two Continents is very similar. When European colonists settled in the eastern States the Indians were already trapping beavers, but they worked individual areas on a conservationist basis and did not seriously affect numbers. Beaver skins later became one of the most important items of trade and great numbers were brought from the native trappers, but it was the white trapper who hit the beavers so hard that they began to lose ground: at the same time forest land had been steadily going over to cultivation, and by the middle of last century the beaver had almost been driven out of the eastern States. The authorities then began to take notice and the State of Maine gave legal protection to this interesting beast in 1866. New York State and Pennsylvania later did the same and the beaver has been re-established so completely that regular trapping—of carefully controlled numbers—is now allowed over considerable areas. A positive programme of conservation includes the distribution of surplus livestock to new areas or to rivers where it has been lost: its economic value is fully recognized and the beaver is not now in any real danger.

The passenger pigeon is reckoned to have been more numerous than any other bird that North America has ever known. From the earliest days of French and British occupation these pigeons were being slaughtered, but in 1878 it was still possible for over 1,500,000 birds to be killed in one nesting area only—in Ontario. They nested in Canada and the eastern States, wintering in the southern States, and their migration movements were often in flocks that obscured the sun for several hours at a time. Such numbers quite defeat the imagination today—except perhaps of those who know locusts at first hand. After 1880 the passenger pigeon began to decline rapidly and soon after 1900 it had disappeared for ever, exterminated by the combination of a number of factors. The clearing of the virgin forests, which made the birds nest in low trees, vulnerably exposed; the development of a net able to take 3,000 birds at a time; the rapid growth of communications which allowed hunters to get quick news of flock movements and follow them up; a series of natural disasters, including sleet storms and forest fires in the nesting grounds: these were some of the external factors, while the passenger pigeon had its own inherent weaknesses—unsuspiciousness and stupidity, a slow reproduction rate and an insistence on flying at low altitudes in vast flocks. For some animals, especially gregarious ones, there seems to be a critical level in numbers from below which there can be no recovery; this pigeon was ultra-social and it had perhaps already dropped below the safety level before man tried in vain to save it.

The beautiful egrets came on the danger list when it became the fashion for ladies to wear the graceful plumes from the breeding plumage; this meant shooting the parents at their nests while the plumes were still unsoiled—that is, long before they had a chance to rear any young. Unrestricted slaughter went on until 1900, when protection was given by the Audubon Society, against such opposition that several of the watchers were killed. The birds were later made quite safe by the legal prohibition of any trade in their feathers, with the result that the snowy and American egrets rapidly recovered.

RESULTS OF COMPETITION AND CONFLICT 39

It was neither competition nor conflict but just a matter of human greed which the public conscience was able to deal with in time. This is quite a good illustration of one of the cardinal principles of wild life protection: if the traffic which threatens the safety of a species can be made utterly unprofitable it is almost certain to stop.

The musk ox is a beast that seemed already on the down grade when man met it. Fossils show that in ancient times it ranged through North Africa and Europe, including France and southern England, but it was driven north by the changing climate and had a much reduced range in Arctic North America before man began to attack it.

It is a strange animal, put in a group all on its own, cowlike in some ways but with horns more reminiscent of an African buffalo and a long thick double coat guaranteed to keep out the cold. Never very common in historic times, it began to decline early in this century. It disappeared from Alaska but was reintroduced there in 1930 and it bred a few years later. Canada meanwhile had made a sanctuary northeast of the Great Slave Lake in 1927 to include the last herd on the mainland. Meanwhile in Greenland the numbers had been heavily reduced, especially by hunters who killed them for their meat, some of which was even exported to Scandinavia. The musk ox is rather stupid; it goes about in small herds and has the habit of forming a solid phalanx facing the enemy—a habit that proves disastrous when applied to an enemy armed with a rifle. Every effort is now being made to save the animal, for its extermination would mean the loss of a quite unique form. In 1951 the Danish Government put into effect legislation protecting the musk ox in north-eastern Greenland and the outlook is much more hopeful.

Several other American animals are on or not far from the danger list, but it is good to know that wild life now has many friends always on the lookout for species threatened with extinction. Action was taken fairly recently on behalf of the trumpeter swan, the largest of the North American waterfowl, with a weight of thirty to thirty-five pounds. Protection was given them in their breeding grounds, and while numbers were still very low they were helped through several difficult periods caused by the failure of the rains and the drying-up of their lakes. Such a bird is specially vulnerable, for its large size makes it an obvious and easy target as well as limiting the waters it can use with safety; while migratory habits may mean passing through regions where protection cannot be enforced.

The tiny key deer of Florida is an example of an animal threatened with extermination largely because of its very limited range. It is the smallest of several varieties of the white-tail deer found in many parts of North America, but it is quite distinct from the usual form and some zoologists consider it a separate species. Formerly found over the Lower Florida Keys—a complex of low islands set in mangrove swamps and brackish water—these tiny deer managed to survive in competition with the Indian inhabitants, but today they are reduced to a mere handful. Robert P. Allen, writing in *Natural History* (Feb. 1951), suggests that there may be no more than thirty specimens of what he rightly calls "these toy deer" alive today (a buck weighs something like 50 lb. and a doe 10 or 15 lb. less). He blames shooting, hunting with dogs, fast cars, forest fires and building development for this sorry state, but he considers that if the measures now being taken for their protection are applied effectively the key deer can still be saved.

4. IN OTHER LANDS AND AT SEA

THE expression "as dead as the dodo" has long suggested something that has gone far beyond any possible recall; with the last dodo died not just an individual but a whole species, with no close relative surviving. It was flightless, clumsy and defenceless and, being as big as a swan, it just could not compete when men arrived on its rather small island home of Mauritius. The name comes from the Portuguese word "doudo", meaning "a fool", which shows just what the sailors thought of it. Admiral van Neck discovered it in 1598; it remained abundant for a time, but had been lost for good before the end of the 17th century, leaving behind a few stuffed skins, some bones dug up from the peat and some pictures which give a good idea of what it looked like in the flesh. The solitaire was another giant member of the pigeon family —also flightless—found on the island of Rodriguez; this also failed to stand up to the competition of man and the various animals he brought to the island with him, and it disappeared a few decades after the dodo.

On the other side of the southern hemisphere, in New Zealand, another group of giant flightless birds was meeting the same fate at the hands of the Maori colonists. Here it is much harder to suggest actual dates, but it is generally thought that the big moas disappeared about the 16th century. There were many different kinds, ranging from about turkey size to something like twelve feet tall, and they are known from the remains of eggshells, skins, bones and feathers that have been found on old camp sites. There have always been some authorities who thought that moas existed until more or less recent years: the re-discovery of the *Notornis* after it had been considered extinct for a century suggests that other birds may await discovery or re-discovery—including perhaps one of the smaller moas.

Steller's sea cow was another animal quite unfitted to compete with man. It seems to have gone for good within about thirty years of being discovered by Bering and Steller when they were wrecked on what is now known as Bering Island. It was a colossal beast, some 20 to 25 feet long and weighing perhaps four tons. It was a vegetarian that specialized on seaweeds, and the sailors who saw it living around the Commander Islands in the Bering Sea described it as stupid, sluggish and comparatively helpless. It was thus an easy prey and, as a result of mass slaughter, it was exterminated in the latter part of the 18th century. Its loss is the more deplorable because it was quite unique and lived a very different life from its nearest living relatives, the manatees of West Africa and Central and South America, and the dugong, found on the shores of the Indian Ocean and East Africa round to Australia. All of these live in comparatively warm waters and feed largely, if not entirely, on vegetation quite different from seaweed.

The rhinoceroses, too, might be said to be poorly equipped for survival in a mechanized world. With their great size and thick skins they have comparatively few natural enemies, and they could also compete fairly well with man until he invented powerful precision firearms. They survive today because they are deliberately protected, but their future is still rather uncertain. The three Asiatic rhinos, found from Sumatra in the south up through Malaya and Burma into Assam, are in danger principally for one reason: because their horns are credited with semi-magical properties and fetch fantastic prices in the East. These horns are often worth more than their weight in gold! These rhinos live in inaccessible jungles difficult to control properly and the rather unstable political conditions, following a period of war with little or no control in the jungles, make the outlook black. The two African species —the black and the so-called white rhinos—are more plentiful and, for the time being, seem safe enough. The former is widespread and often common. The latter, which in prehistoric times was widely distributed and numerous, is now found in two areas only, but it is unfair to blame man for this tremendous reduction in range; it occurred mostly long before he appeared on the scene. In south-east Africa there is a herd of some 200 in Zululand and, several thousand miles to the north, another group in the Lado enclave. Their position is the same as that of many other large animals of the African plains; under the protection by the European powers they are unlikely to be wiped out, but with the increasing pressure on the land and with the probability of self-government by peoples with little or no appreciation of their heritage of wild life, the further outlook is rather unsettled.

The larger the animal, the more vulnerable it is to man armed with modern weapons, but with some groups man does no even need precision arms to do the damage. This is very much the case with the giant tortoises—and they really are giant, for an old specimen may weigh over six cwt. At the end of the Tertiary period, something over one million years ago, giant tortoises were found in many parts of the world and their fossilized remains are well known. Most had disappeared before our era and man seems to have found them living on four island groups, three in the Indian Ocean and one in the South Pacific. The development of sea-borne traffic in the 16th and 17th centuries was the start of the giant tortoise's trouble, particularly in the Indian Ocean. Sailors found them in great quantity and even described them as forming solid packs of 3,000 or more together. In the days before refrigerators they were a marvellous source of fresh meat, for they would remain alive for ages on board ship, and they were freely used for this purpose as well as being exported as meat to other islands. At the same time the giant tortoises were taken from one island to another and the whole lot got thoroughly mixed; there were once about twenty different kinds, but some of these have certainly been lost. The Seychelles species, for instance, became extinct in its native land, but it had been taken to Mauritius and become established, though only to become lost for good at the end of the 19th century. Meanwhile the North Aldabra tortoise went to the Seychelles, which is now its only home, though the South Aldabra form remains safe in its original land. Latest reports are most encouraging and it seems that several species at least should survive.

The Galapagos Islands, in the South Pacific, where Charles Darwin did so much of his original work, is the other home of giant tortoises, and almost each island at one time had its own peculiar type, as was the case, too, with many other animals. The present position there is not known, but it seems rather unlikely that more than one or perhaps two forms of the Galapagos giant tortoises can survive much longer.

Australia and New Zealand are well known for the quite unique nature of their fauna and the devastating effects of introducing a mass of foreign animals, but man himself has also done a frightful amount of direct damage, especially to the Australian mammals. By the middle of the 19th century New South Wales alone had some 10,000,000 cattle and sheep, and the colonists had to make room for this army: the large vegetarian marsupials—the kangaroos and wallabies—were regarded as direct competitors for the grazing and were shot in great numbers. At times the Government put a price on their heads to encourage settlers to exterminate them, but an even bigger factor was the demand for wallaby and kangaroo furs. Exact figures are not easy to arrive at, but in 1913 one and a quarter million wallaby skins went through the London fur sales alone. Men made their living by killing for the fur trade: this would not be difficult in areas where, in 1923, a man was known to get over 2,000 skins on a three months' trip.

It was not only the wallaby group that suffered. Even though they were in no way competitors the opossums were killed, literally, in millions. As far back as 1873 it is recorded that 300,000 opossum furs were sold in London alone; in 1893 the numbers had risen to 2,000,000 and by 1903 to over 3,000,000. These opossums must have been incredibly plentiful, for in the early 1920s America alone was still getting some 2,000,000 of their skins every year. The delightful koala was attacked just as mercilessly. In spite of nominal protection it was shot and marketed as "wombat", and it was under this name that 2,000,000 skins were sold in 1924. Even in 1927 it is reckoned that 600,000 koalas were killed by licensed trappers.

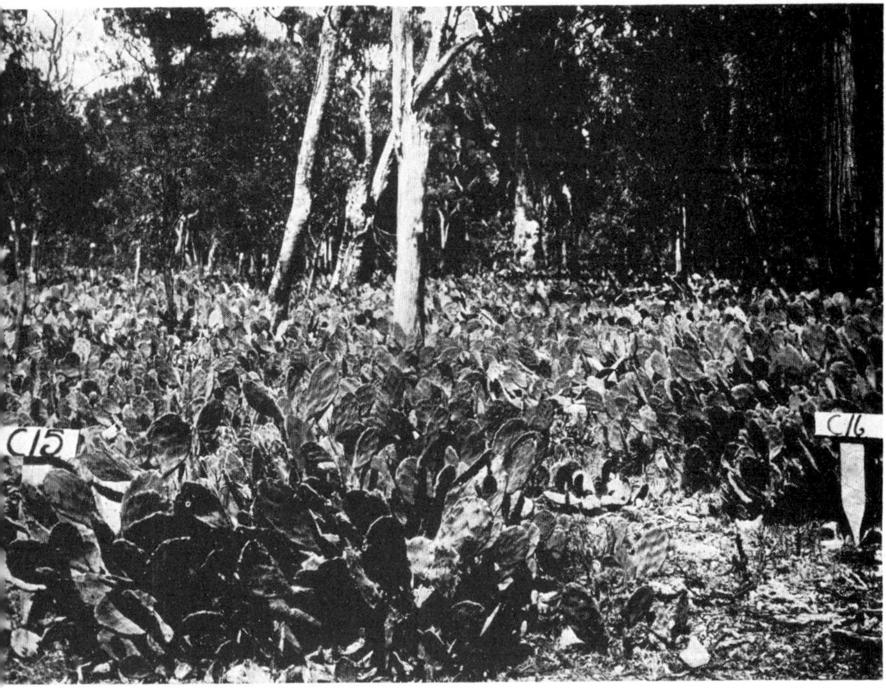

Australian News & Information Bureau

Dense prickly pear in Queensland, Australia

The same area three years later after destruction by *Cactoblastis*

Australian News & Information Bureau

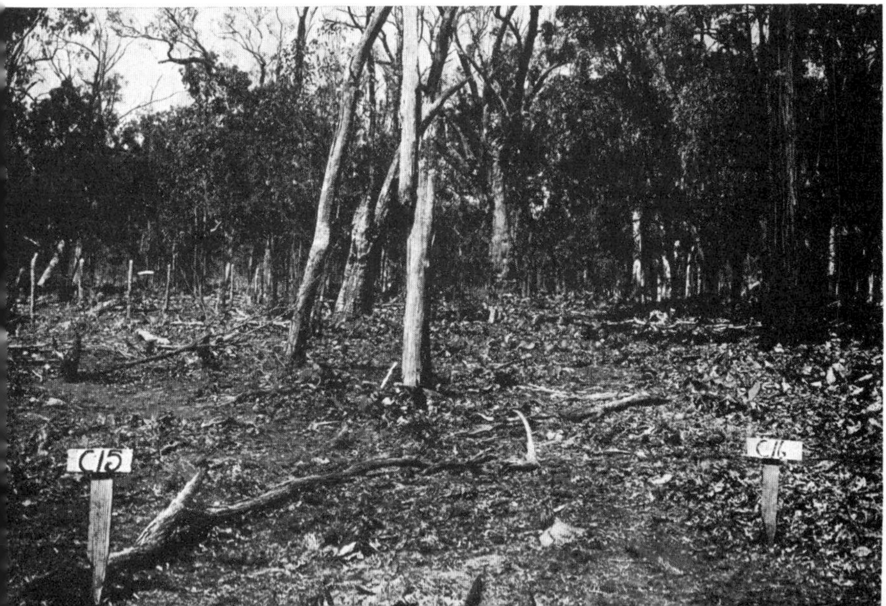

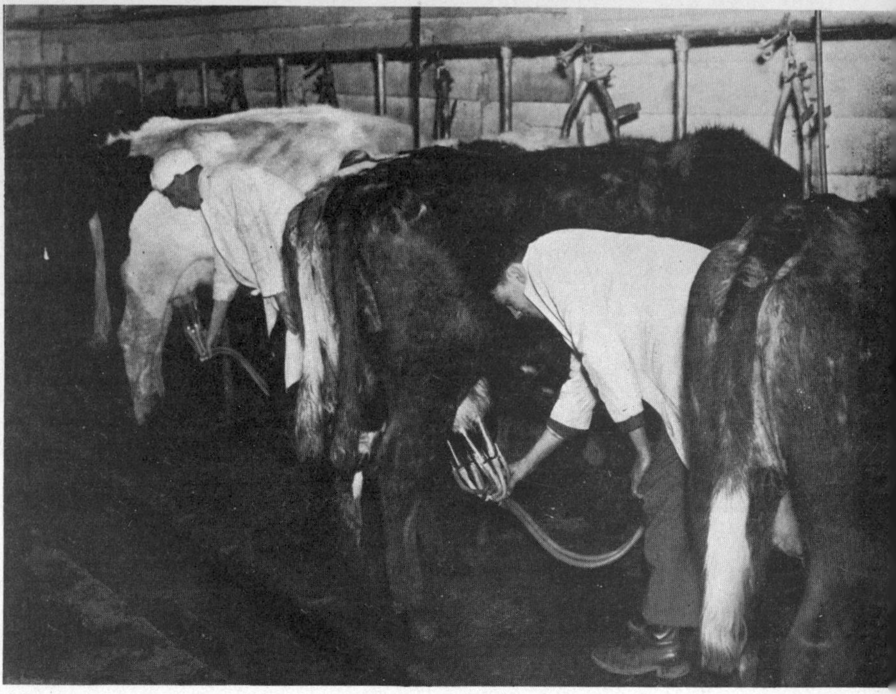

Fox Photos

Milking by electricity

Green-fly are the cows of the ant world

Picture Post Library

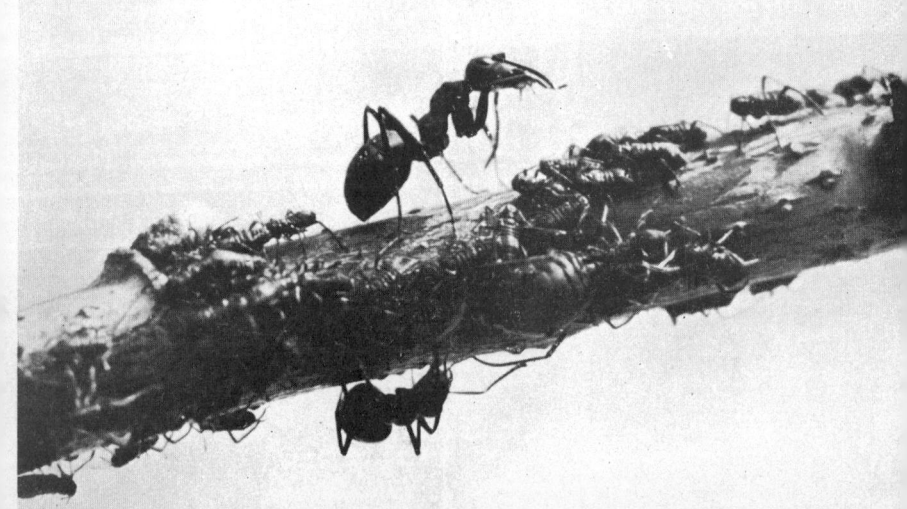

It is now a little difficult to apportion immediate responsibility for the sorry condition of many of Australia's interesting animals, but man is finally to blame because he brought in the rats and rabbits, cats and dogs—and, worst of all, himself. Australia is an immense territory which is still only sparsely inhabited: otherwise the death-roll of species would have been far higher. Over ten years ago it was reckoned that eleven animals were definitely extinct—all except one of these were lost in the present century—but more are on the danger list. Many people have realized the danger and great efforts are being made to save as many species as possible by the formation of sanctuaries and the prohibition of killing. The koala, once so plentiful and well distributed, has disappeared from most of its old breeding grounds and is fairly common only in Queensland. It is now completely protected by law everywhere, but the most important step to this end was taken in the United States of America, where the importation of skins was forbidden. This killed the market and removed the temptation.

The Père David's deer has an absolutely unique history and it owes its existence today to man's direct action. It is a large deer, quite unlike any other known kind, and it was discovered in 1865 in the Imperial Hunting Park a few miles south of Peking by Father David, a member of the Lazarist Mission working in China and the collector of many forms of plant and animal life new to science. Father David saw some of the deer inside this vast walled park and realized that it was something quite new: he was unable to get specimens for the Paris Museum through diplomatic channels and so he arranged things with the park guards. Specimens were obtained in 1866 and the new deer was described formally later that year, but at almost the same time a live pair had been obtained through the French Embassy.

It seemed unlikely that this deer would exist purely as a herd in this park and people began looking for it in the wild, but no trace of it as a living animal has ever been found, though every part of China and Mongolia has been searched. Excavations in the past thirty years have revealed Père David's deer antlers worked on by craftsmen in the Shang-Yin period some 1700–1100 B.C., but after that there is no definite evidence about it until the time of its discovery by Father David. It had lived on the river flats in many parts of China and was essentially a swamp deer: bit by bit its haunts were taken over for rice-growing, and one can only presume that this was the cause of its disappearance, but there is nothing to show just when a herd was enclosed in the Imperial Park. There is thus a gap of about 3,500 years which, so far, Chinese literature has failed to fill.

In the few years after its discovery a number of specimens were sent to European Zoos and this was most fortunate, for in 1894 a flood breached the park wall and the herd escaped, only to be slaughtered and eaten within a very few years. The Duke of Bedford had obtained a pair from the Paris Zoo towards the end of the 19th century and sixteen more were obtained in the next few years from various Continental Zoos: from these was built up the magnificent herd of over 250 that now lives in Woburn Park. The present Duke has sent breeding pairs of this unique deer to parts of the Continent as well as to South Africa, North America and Australia, and as a species it should be safe, even though captive.[1]

The story of the American bison given in another chapter is probably fairly familiar, but comparatively few people even know of the existence of the European bison, or wisent, far less of its very near escape from extermination in two world wars. Although closely related to the American bison and similar in many ways, the wisent differs in several essentials. It is taller, longer and distinctly narrower in the beam: the head is smaller and the head hair less dense so that the head and shoulders look much less heavy altogether. It also differs in its habits, for it is a browser in the forest rather than a grazer on the plains, which is what the American bison is today.

The bison was once found in many parts of Europe and Caesar mentions it as plentiful, along with the aurochs, or wild ox, in the forests of Germany; it was certainly known in the Carpathians and in parts of Germany as late as the 18th century. In most parts it was the same old story—the bison had to retreat in the face of direct attacks by man and a continuous

---

[1] A full account of this deer, from which some details are quoted here, is contained in a symposium edited by Professor F. Wood Jones in the *Proceedings of the Zoological Society of London*, Vol. 121.

restriction of breeding and feeding grounds. But in parts of eastern and south-eastern Europe it managed to find sanctuary in the dense forests, often assisted by direct protection. The Caucasus Mountains had a bison more or less distinct from the main form and the Russian authorities made great efforts to save it: all seemed to be going well until the 1914–18 war, after which there were few survivors, and in spite of renewed efforts the last of these died about 1930, but a little Caucasian blood still runs in some cross-bred animals in several zoos.

In 1914 the typical European bison had two main centres —the forest of Bialowies in Lithuania, which had been the private hunting preserve of the Czar, and the forest of Pless in Upper Silesia. In the former the herd lived at liberty and had reached a peak of some 1,600 head in 1860, but had dropped to about 600 before the First World War. The latter herd had been founded by importing three animals from Bialowies and had grown to over seventy. Both herds were decimated in the war and in the immediate post-war period and had completely disappeared. Fortunately a few pure-bred animals had survived in zoos and three of these were got together in 1923 to found a new Pless herd, and a further three in 1929 to form a new Bialowies herd in Poland. Meanwhile a studbook had been opened and every beast recorded. By 1939 these two herds had reached a total of over thirty. In spite of regrettable incidents in which Allied occupation troops and released prisoners were as guilty as the fighting armies, these herds just survived the war years, and the latest published list, for the 1st January, 1947, includes over 100 pure-bred wisents in eight countries of Europe, including Great Britain. Many of these are kept under more or less zoo conditions and others are semi-wild in large parks. The wisent is thus extinct as a truly wild animal, but should survive long in its present state.

Most of the animals so far discussed live on land and can, therefore, be protected, at least nominally, by one or more Governments. But the marine animals pose a completely different problem, which is more or less stated in connection with the question of over-fishing in the non-territorial waters around the British coasts. For several reasons the position of the big sea mammals is much more acute than that of most fish: they are generally so much bigger that they can be hunted individually: they are much more valuable prizes, largely because of their fat or fur: the products from them can be brought from the far ends of the earth in good condition: finally—though this refers only to the seals and their relatives—they may haul out for breeding purposes in very limited areas and at definite seasons, thus laying themselves open to concentrated attack, but at the same time giving the landowner the chance of protecting them.

The marine mammals are mostly the subject of a sorry history of over-exploitation, with a few bright spots here and there to show what might be done—in some cases, alas, what *might have been done*—if action were taken in time. With few exceptions the fur seals of the southern oceans have been reduced to a tiny fraction of their former populations: these were the animals that once yielded hundreds of thousands of valuable skins from such areas as the Falkland Islands, the far south of South America and Australasia. Bertram cites the case of the South Shetland Islands as typical of 18th- and 19th-century mining of natural resources: their animal riches were discovered in 1820: in the following two years the fantastic total of 320,000 skins were taken—but within ten years the colony was wiped out. It took fifty years for another and much smaller colony to be built up, and the hunters then wiped that out. This sort of thing happened almost everywhere and a wonderful source of raw materials was lost. The solitary known exception in this tale of slaughter concerns a small rookery of a local form of the southern fur seal living on Lobos Island off the Uruguay coast: this is managed on what might be called a "constant yield" basis, bringing a regular revenue to Government and a fairly certain future for the seals.

Fur seals also live in northern waters: they had fewer breeding centres (principally in the Kurile, Commander and Pribilof Islands), but the numbers breeding on some islands were almost astronomical. These breeding grounds were discovered at various times from the end of the 18th century to about the end of the last century, and the attacks began soon after discovery. The Pribilof Islands were under Russian control at first and there was a real attempt to conserve the herds by taking only a reasonable number of skins each year.

W. Suschitzky

A family of guinea-pigs in the London Zoo

This goat mascot of the Royal Welch Fusiliers was not keen to go overseas

P.A.-Reuter

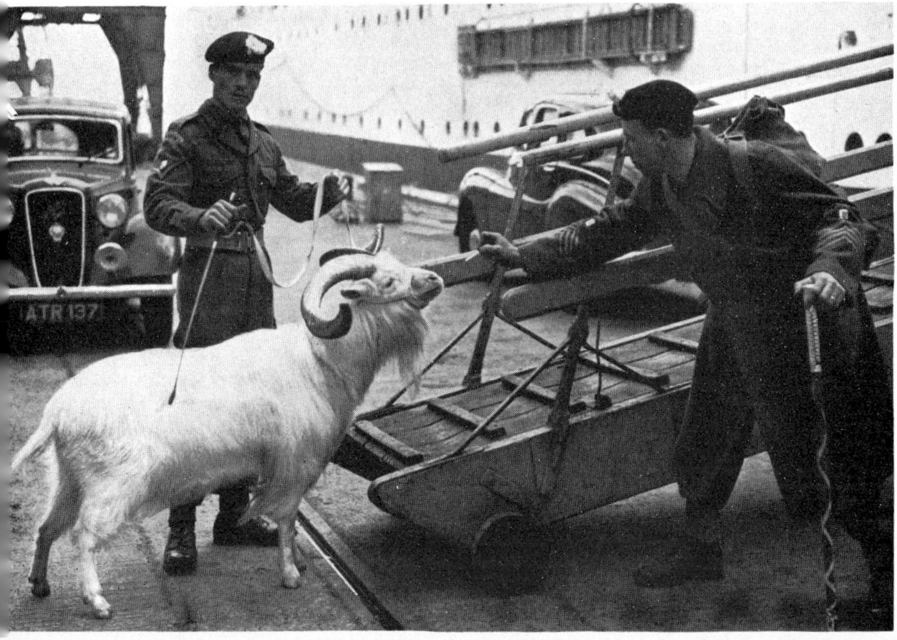

Paul Popper

There is a big contrast between this wild sow and its offspring
and
this domestic pig's family

Farmer and Stock-Breeder

This policy continued when the United States of America bought Alaska and these islands in 1867, but soon a new sort of hunting was practised. This was known as pelagic sealing, for the ships lay offshore, beyond territorial limits, and shot the seals as they travelled to their feeding grounds. This had a drastic effect on the colonies, and by 1911 the total on these islands had dropped to almost 130,000; this was perhaps one-twentieth of what it had been. In that year Great Britain, the United States of America and Japan made an agreement to stop pelagic sealing and the herds have increased steadily to some two and a half millions. Very large numbers are killed every year on the islands, but only three-year-old males are taken: this does no damage to the herds and it merely reduces the competition in communities where harems average sixty and a really successful bull may have 100 cows. It is rather ironical that the latest rookeries to be found—on the Kurile Islands in 1881—were the hardest hit. There were 20,000 or so fur seals to start with, but they had been virtually exterminated by 1898. Today that particular race of the northern fur seal is probably extinct.

The delightful sea otter must be mentioned here, for it is near to the northern fur seal in both habits and history. It is just about the largest of the otters, with a total length of some five feet and a weight of five stone. Its home was in the North Pacific, especially around the islands of the Bering Sea and the coast of Alaska. The Russians discovered it 200 years ago and its skin soon became one of the most valuable on the market, fetching perhaps 1,000 dollars. In 1800 the sea otters were still fairly plentiful and one dealer could still handle 15,000 skins in a year, but by 1826 it was reported that fifteen were being taken where formerly the annual yield was a thousand. The Russians then managed to exert some control in hunting, and the sea otters picked up a bit, but when Alaska passed into American hands uncontrolled hunting began again. It was not until 1910 that the United States of America passed really effective laws to protect the otters; the latest estimate is that there are at least 5,000 of them and a small amount of licensed hunting is now possible. The Russians, too, are giving them protection and are also studying them scientifically as a potential fur-farm animal. So the sea otter should now be safe, for which we should be very glad, for it is a charming animal with more than an average share of the playful ways and *joie de vivre* that characterize the family.

The true seals—which include the common seal and grey seal of the British coasts—have no valuable coat, and without this price on their heads most kinds have been less heavily hunted. Even so, the monk seals of the warm temperate and tropical zones have been heavily reduced and at least three forms, the Hawaian, the West Indian and the Mediterranean, may even be extinct. The colossal elephant seals—up to 25 feet long, weighing several tons and yielding perhaps 200 gallons of oil—had a number of breeding stations in the far south, with another form frequenting some islands off the south Californian coast. Some rookeries have long since been deserted, thanks to the greedy whalers in their quest for blubber; others have a tiny remnant, perhaps a dozen where there were thousands. In a few places, notably in South Georgia near the Falkland Islands, it is exploited rationally in somewhat the same way as the northern fur seal, for it has the same polygynous habits: large numbers of the gigantic adult bulls are killed annually and turned into fat and other useful materials.

The whales, staying at sea all the time, form populations which are much harder to gauge. Whaling started something like a thousand years ago and has become increasingly important down the ages. First the accessible whales were hunted, and then the progressively more inaccessible and difficult varieties as the others dropped to a point that made hunting unprofitable. At first they were hunted for the whalebone, and oil for lighting purposes; as better illuminants came on the market, whale oil went to soap-making and the manufacture of margarine. Today the whales are being turned into oil, meal, bone-meal and fertilizer, with virtually no waste, and an international agreement is in operation which may save the whales from extinction. A total annual catch is agreed on, at present at about two-thirds of the average pre-war figure, and this is apportioned among the nations concerned. Some kinds have been heavily hit, and modern whalers, using fast craft, radar and all kinds of devices, are devastatingly efficient, but if this agreement can run for a few years there will be distinctly more hope for the world's whales.

Part Two

The Results of Introducing Foreign Animals

1. THE DAMAGE DONE

THE opening chapters gave some idea of the devastating damage that man can do to the natural fauna when he takes possession of a country. Not all of this damage was, or is, deliberate, but it all results from his direct action in competition and in conflict with the animals, especially the larger ones: in many cases there was plain, uncontrolled greed. In addition to this, man has caused great and far-reaching damage indirectly, by introducing alien animals. (He has introduced alien plants as well, but that is quite another story.) In some cases, particularly on islands where there were once many unique forms, the indirect damage has been even worse than the direct. The first few introductions, such as rats and mice, were presumably accidental (apart from the special case of domesticated stock) and resulted from carrying the animals as stowaways or among baggage in ships. Much later there came a period when a wide variety of animals was introduced into many countries, sometimes as a potential source of meat, sometimes to control other animals, but often just to remind colonists of their far-away homeland. This period is now past and any introductions that occur today are either stowaways that escape detection or thoroughly vetted species brought in for special purposes.

The people who introduced the foreign animals that afterwards became large-scale pests did not intend to create a nuisance: they just did not realize that they were playing with fire! Natural history is an old subject but it is only in fairly recent times that we have really studied the relations between the animals on any area and the plants, soil and climate around them, as well as between the various animals themselves; this is known as the science of ecology and it is a fascinating business. It was the failure to realize how intricate these relationships often are that led to the damage.

In any area reasonably undisturbed by man the natural animals and plants have reached a fairly balanced condition in which predators and prey more or less hold their own, though nature is never absolutely static and there may be quite big variations in numbers and kinds of animals as well as plants. The introduction of some foreign animals into a balanced community may be very serious: a predator, such as the mongoose taken to the West Indies, may wreak havoc on animals which have never learnt to defend themselves against its particular line of attack, while another newcomer, finding an unoccupied niche, with ideal conditions and few or no natural enemies, may get right out of hand and become a serious plague, like the rabbit in Australia. The more specialized the animal, the more difficulty it has in getting established and the less likely it is to become a pest. Also the bigger it is and the more desirable for human food, the easier it should be to keep in check, provided that the human population is sufficiently dense: the pheasant in Great Britain is a good example of this.

The very first of these chance introductions must have taken place back in man's early days—as soon, in fact, as he began to move about to any extent—but it was probably not until sea transport developed that any major mischief was done by bringing new, often aggressive, elements to lands from which natural obstacles had previously excluded them. For instance, the first human invaders of Great Britain, long before the Romans, brought sheep and goats with them, and these were probably one of the important factors in clearing the hilltops in southern England and forming the open downs of the chalk and limestone areas.

Comparatively few mammals have been classed as chance introductions; indeed, rats and mice appear to be the only ones of importance. The black rat, or house rat, came to Europe in early times from Asia, and by the Middle Ages it had already spread over most of the Old World, including Great Britain. The brown rat came much more recently—at the beginning of the 18th century—but proved more successful than the black rat, which it has largely replaced in many areas. Fraser Darling draws attention to the damage done by rats in the Scottish islands. They were even responsible for the desertion of North Rona by the human population of about thirty towards the end of the 17th century. These were black rats and they came from a wrecked ship; they took complete possession of the island and ate up all the food stores, with the result that the people starved to death. Then there was nothing more for the rats to eat and the seas were too rough for them to hunt the tideline, so they died out too. On other islands the brown rats live a scavenging life on the shore in winter and raid terns' colonies in summer, sometimes exterminating them: but the herring gulls' eggs are a bit too big and so they are left untouched.

The house mouse, an even more intimate associate of the human race, also came from Asia, and is now world-wide in its distribution. Unmitigated pests everywhere, these rodents were quite certainly not taken of his own will by man when he moved to new countries, even though it is clear that the black rat and the house mouse arrived in North America very soon after the first European visitor.

Rats and mice are now so firmly established in and around human habitations that they seem unlikely ever to be exterminated, though certain specialized poisons discovered in recent years make the position look rather more hopeful.

Practically all the chance introductions, both early and recent, have turned out to be very unwelcome guests in the countries of their adoption. One such invader has recently come to light in North America, and it is an interesting case of indirect human influence. The Welland Canal was constructed to by-pass the Niagara Falls and make the principal Great Lakes accessible to ocean-going shipping, but at the same time it has allowed the lamprey to get up into the lakes and attack the trout there. These lampreys formerly lived in salt water, but they gradually moved up into Lake Ontario and thence, either by their own power or attached to ships, to Lake Erie. By 1948 the catch of lake trout in the United States section of Lake Huron had dropped from an average of nearly 2,000,000 lb. per annum to under 5,000 lb., and scarcely a trout was without sores or scars caused by the blood-sucking lampreys. Biologists are still far from finding an answer to this scourge which threatens to exterminate a very important game and food fish.

In marked contrast is a little nocturnal lizard—Brooke's gecko—which apparently has the distinction of being the only reptile (perhaps the only land vertebrate) to be found right through the tropics of both Old and New Worlds. We cannot be absolutely certain, but it very much looks as if this wide distribution took place with unwitting human assistance. Brooke's gecko is one of the most enterprising of the lizard family. Long ago it found that human habitations, lighted by night, made excellent hunting grounds, and few houses in Central Africa are without some of them sleeping in the cracks in the walls or behind pictures by day and taking station around the lamps by night to feed on the hosts of insects attracted to light. It was therefore in a very good position to be picked up and carried around by man in his goods and baggage. It is likely that Brooke's gecko first lived in Africa, but now it has travelled widely and is common in Central America and Asia as well. Its diet is restricted to insects and it does a lot of good, even though it may take a few useful insects. There would be no complaint if all animal hitch-hikers belonged to this class.

Before World War II, when bananas arrived by the shipload, the stems sheltered quite a variety of stowaways—spiders, centipedes and lizards, especially geckos—most of which found their way to the Zoo if they were discovered alive. Even now we get an interesting lot of specimens that have come to England in this way. Bundles of esparto grass brought from North Africa for paper-making are often found to hold such treasures as scorpions and lizards, and one consignment of camel bones from Egypt yielded a desert skink.

Most of the other chance introductions belong to the lower animals, generally referred to as invertebrates, or animals without backbones. Without exception they seem to be unwelcome. The *Anopheles* mosquito, carrier of malaria, jumped a lift across the Atlantic from West Africa to Brazil, and the successful campaign waged against it gives some idea of what such an introduction can cost. These details are taken from official reports. A flourishing colony of the mosquito was discovered in irrigation ditches near Natal in March 1930, and it was already found to have spread some 200 miles inland and 220 miles along the Brazilian coast. It carried the dangerous form of malaria known as malignant tertian: in one large valley practically all the inhabitants contracted it and the death-roll was estimated at 20,000. Two hundred and sixty tons of arsenical preparations were used to kill the mosquito larvae and 6,000,000 tablets of atebrin and quinine were used. The battle lasted for twelve years and at times the staff numbered 3,700; it cost no less than 2,000,000 dollars to wipe out that mosquito.

New Zealand also learnt expensively that insects can travel in aircraft. In 1945 the common European wasp arrived on North Island and, in spite of a campaign with D.D.T. and cyanide, wasps became established over an area of some 200 miles by 150 miles. Eradication is now considered to be quite impossible and the only hope is to keep the wasps away from the fruit districts. Very strict fumigation laws are now applied to aircraft in many countries, with the result that mosquitoes, tsetse flies and other insect pests and carriers of disease are unlikely to get through alive.

The Colorado beetle is a pest that has crossed the Atlantic from the southern United States and has been much in the news recently. In the wild both grubs and beetles feed on the leaves of plants belonging to the family *Solanaceae*, to which the tomato, potato and deadly nightshade belong, and crops of potatoes provide conditions in which they can multiply at great speed. The beetle's first appearance in England was at Tilbury in 1901, a chance arrival by sea, but it was not until 1933 that it was seen again, also at Tilbury. Meanwhile it had become established in great numbers in parts of Europe, including the north coast of France, from which it can easily hop aboard a cross-Channel steamer and also be carried over by suitable winds. Small numbers have been found on crops in Great Britain from 1941 onwards, but so far it has not become fully established and, with continual vigilance, it may still be possible to keep it out.

At first sight the crustaceans would seem an unlikely group in which to find world-wide hitch-hikers, but they include at least one serious pest—the mitten crab. It is a native of the river estuaries in northern China, but by 1912 it had travelled to north-west Europe, and it was found in one of the tributaries of the River Weser: presumably it had travelled in a steamer's ballast tanks. By 1933 it had spread over most of North Germany and Holland, and it was known in Belgium: it had reached beyond Königsberg on the Baltic coast: and it had travelled up the Elbe as far as Prague and to beyond Mainz on the Rhine. It makes itself a serious nuisance in various ways: it damages banks, and this may be very dangerous in the case of canals above land level, but it is perhaps most annoying to the fishermen, for it actually eats their nets as well as filling them up—as much as half a ton of mitten crabs have been taken in one net on the lower Elbe. Like most salt-water crabs it is carnivorous, so it eats the fishermen's baits and generally competes with the more valuable fish in the sea. Control seems to be almost impossible, and the mitten crab is likely to stay in Europe.

The Mollusca is another order that has provided a pest much in the news just after the recent war—the giant snail of East Africa. It arrived in Ceylon and India by easy stages and has thence spread to Malaya, Indo-China and the Pacific Islands, partly carried unintentionally, partly a deliberate introduction by Chinese or Japanese as food for their ducks, though only the very young stages could possibly be eaten by any ordinary duck, for it grows to a length of several inches! Details of the snail's wanderings have recently been written up and in most places it seems to have become a plague within a few years of arrival. The strongest measures have been enforced to stop further spread, and the penalties threatened for taking a live snail into the U.S.A. are quite frightening. All possible control action has been taken in areas already overrun and scientists are busy looking for natural enemies in its homeland. It also seems as if local predators, especially some beetle larvae, are likely to attack the snail and in some areas, at least, the seriousness of the invasion decreases after a few years. The danger may therefore not be quite so great as it seemed at first. A closely related giant snail is found in West

RESULTS OF INTRODUCING FOREIGN ANIMALS

Africa, where it is much used for food; so much, in fact, that collection is allowed only every second or third year and the stock is thus conserved. It is a pity that man does not have the same tastes in the Far East, but without the conservation part.

When we consider man's deliberate introductions of wild animals, most of which took place in the 19th century, it is impossible to deny that, with a few exceptions, these introductions have turned out to be tragic mistakes. In Great Britain we have a number of examples, of which the grey squirrel is the most notorious. It seems to be more or less omnivorous and, finding itself without the large birds and beasts of prey that control it in its native North America, this squirrel has spread over a great part of England from its first effective introduction in 1876. At least thirty specimens were imported between then and 1910, and man helped the grey squirrel further by distributing the progeny of the original stock. It is usually blamed for the decrease of the British red squirrel's numbers, but though it is to some extent a competitor, it is not entirely responsible for the latter's scarcity. It seems likely that the red squirrel was the victim of an epidemic, from which it is now recovering in some areas, and we can only hope for a similar disease in its fairly closely related American cousin. It is rather unfairly called "tree rat" in some parts, but it has the rat's very unwelcome habit of destroying eggs and young birds. The grey squirrel sometimes reaches numbers that are normally associated with rats or other smaller rodents: in the spring and summer of 1951, for instance, over 270 squirrels were shot in the walled garden of a large Surrey house, mostly as they came to raid the strawberries. In spite of such punishment they got at least half of the crop. No wonder the grey squirrel is good eating! Unfortunately the scarcity and cost of cartridges in these days tends to discourage gamekeepers and others from shooting such pests. Opinion is unanimous that the grey squirrel in England should be attacked everywhere, always and by all means: though this point seems not to be generally known, it is, in fact, illegal to keep one alive without official permission.

Another rodent was first introduced shortly after the grey squirrel—the fat dormouse from Continental Europe. It is something the size of a rat, comfortable in proportions and with a very bushy tail. Lord Rothschild released a few pairs in the Tring area in 1886 and there were several subsequent importations. It established itself slowly, but gradually spread into several adjoining counties. At present it seems to be doing little serious damage and keeps largely to the neighbourhood of houses, sometimes being found in roofs, and it is too early to say yet whether it will become a pest. In other parts of the world, especially Africa, dormice are well known for being carnivorous and may do much damage to birds' nests and eggs, but this charge has never been laid against the fat dormouse, which seems a fairly strict vegetarian with a preference for fruit. It is, however, considered potentially dangerous enough for the Ministry of Agriculture and Fisheries to have initiated a survey of the problem. It is also known as the edible dormouse, and the Romans bred it for the table, to be served with a honey sauce. It is recorded that Q. Fulvius Lippenius introduced the practice of breeding it towards the middle of the 1st century B.C. In the northern parts of its range it undergoes a deep hibernation and the Germans call it the "*Siebenschläffer*"—the "Seven-sleeper"—because it nests communally.

The little owl is still the subject of argument. After being introduced from southern Europe in the latter part of the 19th century it gradually made itself at home and must be considered fully naturalized in England, though it has not spread to Scotland or Ireland. As Aristotle noted over 2,000 years ago, the little owl does some of its hunting in the twilight and early dawn, and it is true that it takes a few common birds, especially in the breeding season; but careful investigation has shown conclusively that rodents and large insects make up the bulk of its food and it must therefore be considered as a useful rather than a harmful settler. The general opinion in some of the European countries where it is at home is that it is distinctly useful. There seem to be a few records of the little owl in Great Britain before it was successfully introduced: it is thus on the British list in its own right as a rare vagrant and, introduced artificially, as a resident breeding species.

The pheasant is the only other alien bird of importance now established in Great Britain. It was introduced from south-west Asia—probably by the Romans and certainly before 1059, when it was mentioned, with partridges, in the

Waltham Abbey Ordinance. Whether or not the Romans brought the pheasant with them, they knew this bird well enough at home, and by the end of the 1st century A.D. it could be seen in the farms in Italy. Palladius actually included a chapter on pheasant-keeping in a book on agriculture dated in the 4th century. The birds had been known in Athens in the time of Aristophanes, some 400 years B.C., though they were rare then. The pheasant seems to have become naturalized, though it is doubtful whether it would survive permanently without some human assistance.

Two more gamebirds have been introduced into Great Britain in more recent times. The French or red-legged partridge was first released in Suffolk about 1770 and subsequently in various other places as well, and it is now firmly established over much of England from Yorkshire southwards. But it is thought that it also comes from the Continent occasionally under its own power and so its status is very like that of the little owl. In habits it resembles the common partridge, but it is not reckoned to give such good sport.

The capercailzie has a different story, for it is really a reintroduction: it became extinct in Scotland about 1760 and new stock was brought over from Sweden in 1837. The purist might contend that a different form has been introduced, but the former British birds were just the same as those found throughout coniferous forests of northern and central Europe.

The proposed introduction of the reindeer comes into somewhat the same class as the capercailzie, except that it became lost to Scotland much longer ago, probably about the 12th century. Changes in climate were largely responsible for its going, and since then there has almost certainly been a further slight change to its disadvantage: it may therefore prove difficult or even impossible to re-establish.

As far as most people are concerned there is only one British frog: strictly speaking they are quite correct, for only one is native to this country, but two others have been introduced from the Continent and have become more or less established. One of these, the marsh frog, has the distinction of being the biggest European kind, and this comes from eastern Europe, especially from such areas as the Hungarian marshes. A dozen of these frogs were liberated in Kent in

1935 and they have founded a considerable colony in the Romney Marshes: the males are extremely noisy at night and are most unpopular in the neighbourhood of houses. The second is the well-known edible frog, popular on the table in France but little eaten elsewhere. It is about the same size as the common frog—around three inches long—but keeps much more closely to the water: England is just outside its natural range and in spite of many introductions it survives in only a few places in the south-east counties. A series of favourable years probably lets a small colony develop, but then the fickle British climate steps in and the colony more or less disappears.

It may come as a surprise to many people to find the wild rabbit listed as an animal introduced into Great Britain, but such it certainly is, for it was unknown here until it was brought from Central Europe in the 12th century. Although rabbits provide a considerable volume of meat every year, and quantities of skins, they are on balance a frightful pest and the country would be far better off without them. They have spread almost everywhere, even to many of the islands, and have had far-reaching effects on the vegetation. They prevent the regrowth of forest by nibbling off the seedlings as they appear and vastly increase the cost of new plantations by making necessary the erection of rabbit-proof fences. By stealing valuable grazing they reduce the number of sheep and cattle that can be kept and they let weeds flourish at the expense of grass and corn: they are a curse to market gardeners. Although trappers take a heavy toll they make no attempt to exterminate the rabbits from the areas they work, and they are but an example of vested interest. Fraser Darling notes, however, that in Scotland the abundance of rabbits has helped such animals as buzzards, wild cats and pine martens to come through a difficult period and the buzzard is now welcomed as an ally in the war against the rabbit.

The starling is well known as a nuisance in London and other British cities: it has also found its way, with human aid, to North America, and there it seems an even worse pest. In Toronto, for instance, starlings roost in their countless thousands after spending most of the day outside the city limits: trees are plentiful and they use these rather than buildings, but the noise and the filth they cause are beyond description. All efforts to get rid of them have failed: heavy fireworks have proved useful and as flocks settle again they are again moved on, but this only transfers the plague to somebody else and provokes complaints about babies and invalids being woken by the explosions. Fortunately for the Canadians their winter is too severe for the starlings, which move south towards the end of October—only to make themselves a nuisance in Texas by picking out the warble-fly grub from the backs of cattle and damaging the hides.

After such tales of woe it is pleasant to record one harmless animal whose introduction to fresh pastures has been a good thing. The giant tortoises of the Indian Ocean islands are dealt with in more detail in an earlier chapter, and here it need only be noted that if some kinds had not been introduced to various new islands they would have been completely exterminated. This, however, is very much of an exception, and the list of mistakes must be resumed.

Guadaloupe is a tiny mountainous island off the Pacific Coast of Mexico and some 200 miles south of San Diego. It is completely isolated and, as a result, it was at one time well populated with interesting and unique forms of animal and plant life. It was a great breeding ground of the now almost extinct Guadaloupe fur seal and the Russians started visiting the island about the middle of the 18th century to hunt these fur seals, as well as whales and other prey in the Pacific. Fresh food was a problem in those days and perhaps it did not seem unreasonable to release a few goats on the island to provide meat for the whaling-ship crews. L. W. Walker, writing in *Natural History* (Nov. 1947), states that these goats built up to a total of over 40,000 and practically denuded the island of all vegetation: trees were browsed as high as the goats could scramble and no regeneration could take place. Many unique plants vanished for ever, and with them many animals, especially birds: probably many of them had not been described, so we shall never know the total damage done. For sheer destructiveness and ability to tackle almost any kind of vegetation the goat is second to none.

The most striking examples of introduced animals becoming large-scale pests are found in Australia and New Zealand. The introduction of the wild rabbit into Australia will perhaps remain the greatest single mistake ever made in this line, for it has cost that country hundreds of millions of pounds. The first wild rabbits were imported in 1837, but the real menace dates from 1859 when twenty-four rabbits were liberated on an estate near Geelong. In the next six years the owner of this property killed 20,000 but had to admit that there were nearly as many still left. As far back as the 'eighties the Government of New South Wales spent £1,500,000 in seven years trying to control the pest, and that was a lot of money in those days. As in Great Britain, rabbits are fond of other things than grass, and they have cleared enormous areas of tree scrub by barking trees and eating seedlings as they appear. They are also selective feeders in pastures and gradually cause deterioration by letting the unwanted plants take control. There are over 100,000,000 sheep in Australia and it is estimated that there are some three rabbits to each sheep. It has therefore been reckoned that without rabbits the Commonwealth could carry 20 per cent more livestock. It is quite impossible to compute the annual loss to Australia caused by rabbits, but a recent guess is some £40,000,000, quite apart from direct expenditure in pest control. On the credit side must be put the vast numbers of skins and carcasses exported—428,000,000 skins and 90,000,000 carcasses in the five years ending in 1949—but this is very inadequate compensation. The cash value for the year 1948–1949 was nearly £10,000,000.

But the matter does not by any means end there. Intensive poisoning campaigns have done terrible damage to the native birds and mammals. Cats and foxes were imported to deal with the rabbits, but they found the native animals easier prey and generally left the rabbits alone. The foxes tackle most mammals except the larger kangaroos and they chase even these until the young ones fall out of the pouch. There are records of black swans being killed in large numbers by the foxes. The domestic cats, long since reverted to the wild state, concentrate on the native birds, but also attack some of the mammals, and there seems little doubt that the pouched mammals as a whole are not able to stand up to the more highly developed mammals from other parts of the world. The same thing is true to a large extent of the native birds, which find such European immigrants as the starling, sparrow and

South African Railways
Ostriches are still kept on a few South African farms

British Museum

An Egyptian carving, about 2000 B.C., showing a bee sucking honey

Worker bees filling the comb

Eric Hoski

various other finches too enterprising for them. In addition, the starling is a frightful pest in the fruit orchards.

These varied animals which have become very undesirable aliens have done untold damage, but the native species have also had to compete with the armies of cattle, sheep and horses, totalling well over 100,000,000, which man has established. It is little wonder that they have suffered severely.

In New Zealand the position is rather different, for its fauna was at one time quite unique, with only two native mammals, both of them bats. The Maoris had arrived with the Great Fleet in the 14th century, bringing with them the Polynesian dog and a rat native to many of the Pacific Islands: the former was no doubt just a human companion, and it is possible that the rat was used for food. Both were liberated but neither established itself fully, and the dog has quite disappeared. It is hardly surprising, therefore, that the European colonists felt it almost a duty to introduce many of the animals with which they were familiar in their homelands, especially if they were likely to provide good sport. At first this was done in a rather haphazard way, but Acclimatization Societies were soon organized which carried on the work into the present century. Something like 45 kinds of mammals were imported and liberated and of these 25 have established themselves in greater or lesser numbers in the wild state, in addition to the rats and mice which had come uninvited in the ships. They included wallabies and opossums from Australia, wapiti from North America, sambar deer from India and red deer, stoats and cats from Europe. In addition, no less than 130 kinds of birds have been introduced since Captain Cook first landed in October 1769, and it is estimated that 24 of them may be considered established.

Many of the mammals brought in found completely unoccupied niches ideal for them and multiplied rapidly: the pigs and goats—which reverted to the wild form though imported as domestic stock—and several kinds of deer are perhaps the most important elements largely because of their size. For the deer particularly there has been a policy of organized slaughter for many years, and though their numbers have been considerably reduced they are still a pest. In some years more than 100,000 deer have been shot. The rabbit has become a plague nearly as serious as in Australia and, together, these mammals have brought about far-reaching changes in the quite unique forest flora. In addition they have caused serious erosion in some of the mountain areas and in this way, as well as by direct competition, they reduce the numbers of farm stock that can be carried. Several smaller carnivores, notably the weasel, stoat and ferret, were imported as natural enemies of the rodents: they have no doubt exercised some control on these, but they generally found the native birds easier prey and have helped to put some of them on the danger list. The birds introduced to New Zealand were mostly of British species: some are useful or neutral, but others have turned into pests, including the skylark, which is now listed as the second worst bird pest there.

On the other hand, New Zealand provides an excellent example of the successful introduction of sporting fish, especially those of the trout family, into streams poor in native species. Some kinds, especially the salmon, have proved impossible to establish, but the brown and rainbow trout-fishing in many New Zealand rivers is now world-famous.

These then are some of the wild animals (and a few domesticated forms which easily revert to the wild), varying infinitely in size and family, which with man's conscious or unconscious assistance have successfully invaded and made themselves felt in almost every corner of the inhabited world. Man tends to learn lessons slowly and rather painfully, and it was not until well into the present century that he realized that introducing animals into new surroundings was a dangerous practice. Australia and America, two of the worst sufferers, now impose the most stringest controls on the importation of livestock: it seems, in fact, virtually impossible now to import any wild animal into Australia, and even single specimens of birds are refused permission to go through as transit air cargo. In the U.S.A. the prohibition is not so absolute, but mongooses and fruit bats are totally banned lest they should become pests on the chicken farms and fruit farms respectively. In most countries there is provision for very drastic inspection of imported goods, especially of vegetable produce, to prevent the unintentional smuggling of such unwanted pests as the Colorado beetle.

RESULTS OF INTRODUCING FOREIGN ANIMALS

Apart from alien animals deliberately introduced by man and liberated among the natural wild fauna, with results that he may or may not be able to control, there is, of course, the whole range of domestic animals which he has introduced into new countries. In a few cases already mentioned these animals have escaped or been released and have become wild, but in general man retains control of the domestic animals he introduces. In many parts they are, moreover, dependent on him for both food and water, especially water.

In previous chapters we have considered when and where animals became domesticated, but there was little mention of this more geographical aspect of man's action in adapting them to his purposes. In very early days there was considerable movement of stock within the Near East countries and the human migrations were, in fact, possible only because of the flocks and herds. Early in the present era civilized peoples spread north and west from the eastern Mediterranean region, again taking their domestic animals with them. In recent centuries this movement has taken place mainly from the temperate zone of western Europe to similar climatic zones of European settlement in North and South America, Australia, New Zealand and South Africa. But it has also taken place to some extent from the temperate to the tropical zones, and within the latter. Here we encounter another aspect of human action affecting the geographical distribution of animals, namely the introduction of domestic animals to suit different climates and the selective breeding of varieties suited to life in those climates.

In temperate zones the modern stock-breeder's job is fairly simple, for he has a wide choice of animals suitable for many different conditions and for all purposes. The expert's task now is to make stock available for other habitats, especially in the tropics, since here lie the greatest areas still relatively unexploited for food production by modern methods. There is likely also to be a future for other specialized animals in the sub-arctic regions of northern Europe.

Perhaps the two most important qualities required in parts of the tropics are heat tolerance and resistance to disease. In Fiji it was found that various breeds had different degrees of tolerance to heat, a quality previously unconsidered, for stamina and high productivity had been the points generally aimed at. This factor is even more important in areas like northern Queensland, where the annual average shade temperature is no less than 84°F., with a summer maximum of 118°; here the introduction of humped Indian zebus from Texas for crossing with Herefords has produced an animal which puts on more flesh and matures earlier than any English pure-breds and has allowed the occupation of 7,500 square miles of previously unused cattle country. This heat resistance is partly hereditary, but hair colour and texture are also important, for a sleek, pale coat makes a much more efficient insulating jacket than one that is dark and rough.

In Africa the problem of disease is generally more important. The normal policy in Colonial territories is to cross suitable beef breeds with the native African zebu and combine the weight and quality of the former with the immunity to some tropical diseases of the latter. The complete answer is far from being found, but it probably lies in a combination of hygiene, medicine, management and selective breeding.

2. CORRECTING THE BALANCE BY BIOLOGICAL CONTROL

We have already seen how man, by introducing foreign elements into the natural fauna of a given region, more often than not dangerously upsets the balance of nature. This balance, in which predators and prey have reached a state of comparative equilibrium, is not a very stable one and animal populations do, in fact, vary tremendously. Some of the more spectacular insects, for example, have great ups and downs, and butterflies are said to have their "years".

All sorts of factors combine to control the numbers of animals. In a very variable climate like that of Great Britain, some animals like the Dartford warbler and the bearded tit, for instance, are brought to the verge of extinction by a few bad winters, but for many animals the climate is not a major controlling factor and other things step in. If, for example, the insects and larger animals that eat vegetation went on reproducing themselves unchecked, in a fairly few generations they would be far too numerous for the available food supplies: incidentally, man himself seems to be reaching this sort of state in some parts of the world. In nature, however, this overcrowding does not often happen: swarms of locusts occasionally get out of hand and devastate considerable areas, and various caterpillars may defoliate the odd oak tree, but except for invasions of plagues of locusts it is relatively rare to see plants completely stripped of their leaves. Various factors generally step in to get the animals under control before their food supplies are endangered: their enemies may increase, or perhaps their reproductive rate may be reduced. In some parts of the northern hemisphere there are fairly definite cycles of numbers in various groups of animals: nobody knows quite why this should happen but the numbers rise to peak and drop off to a minimum in a rhythmic way.

We cannot say how the various factors combine to produce, and very roughly maintain, the *status quo* for any species of animal: often we do not know all of the factors involved. We certainly cannot say what long-term results ensue from a change in existing conditions, however that change may be caused, but we can say without hesitation that man interferes with existing conditions at his peril. When, therefore, man transports animals and plants from one country to another—sometimes deliberately, sometimes by chance, sometimes in spite of every effort not to do so—the results may be disastrous. In the absence of their natural controls such invaders may increase in their new surroundings until they are far more numerous and destructive than they could ever become in their native haunts. They may even acquire entirely new habits and thus present quite new problems.

The principle of biological control is to enrol the various agents which control these immigrant species and generally prevent them from being a serious pest in their homelands, and introduce them to places where their prey has got out of hand; with luck they will then readjust the balance. The object of biological control is to prevent these pests from doing significant damage by keeping their numbers down. It is not necessary actually to exterminate the troublesome species, though if this could be done nobody would grumble; in the ordinary way the predators used in biological control are so dependent on their hosts that if these were killed out the predators, too, would die.

Rather more loosely the term is sometimes applied to measures taken to control animals which tend to become a nuisance in their homelands after the balance there has been upset by human activities. For instance, insectivorous birds help to maintain a balance in plantations, gardens and other areas, which man has tried to develop for his own purposes, and in some places the birds are actively encouraged to assist him. This, in a sense, may be termed "biological control", but the birds' diet is too general for them to be used for very specific jobs, and they take both harmful and useful insects as well as other groups like spiders, which are sometimes as general as insectivorous birds in their feeding. The fact remains, however, that there are various examples on record, especially in Germany, where areas in which birds were definitely protected have escaped insect plagues that seriously damaged neighbouring areas; so it is rather hard to resist the conclusion that on balance those birds do distinctly more good than harm, even though the biological statisticians say that this is quite unproven. A rather similar argument applies to the insect-eating bats which are so numerous in many parts of the tropics: the volumes of guano made is ample evidence of the colossal numbers of insects destroyed, but they inevitably take both good and bad. In some parts of the southern States, notably in Texas, special houses are erected to encourage them to settle and help in mosquito control. A sounder biological means of tackling mosquitoes is to use tiny fish which feed on their larvae: one called the top minnow is now released in suitable parts of the tropics and is proving quite effective.

The ancient Egyptians have some claim to be the first people to practise biological control. They lived in the granary of the ancient world and at times their stocks of grain must have been very large: their cats apparently had the main job of keeping down the rats and mice, by that time thoroughly established as human parasites, that were always a potential menace in the corn stores. The cat was rather slow in spreading into Europe and before they became common in ancient Greece it was the practice to keep polecats in and around houses to deal with the mice: these were probably the forerunners of the domestic ferrets. The same custom was found in Rome, and Petronius makes the amusing comment that if pet birds disappeared it was the habit to blame the polecats!

The domestic cat has developed along various lines since those far-off days, and mousing may no longer be the main reason for keeping them: even so, cats are still probably the biggest single factor in keeping down house mice in Great Britain, and many warehouse cats get their official milk ration for services rendered in this way. Darwin was fond of pointing out one effect of the cat's hunting: the cat kills the mouse that eats the humble bee that fertilizes the clover, with the result that the clover seeds more heavily around farmhouses!

Mice and rats are always present in Zoos, and the food thrown by the visitors often makes trapping difficult. In the London Zoo several birds have been enlisted in what is, essentially, a biological control scheme. The pheasantries have borrowed two laughing jackasses for the job, big Australian kingfishers that enjoy a mouse as much as anything. When first posted there the jackasses found a very innocent mouse population and they had a wonderful time: I visited them soon after they moved in and I saw one of them sitting on its perch so full it could hardly move, with one dead mouse in its beak and another laid neatly on the perch waiting till there was room for it. The mice soon learnt some wisdom and now the kingfishers are moved up and down the range of aviaries, spending a day or two in each. The pheasants just take no notice of them. Some tawny owls are also used in parts of those aviaries where rats can penetrate, but it is safe to put them only with the larger varieties of pheasants. On the other side of the gardens the tropical bird-house had long been a problem, for every effort at mouse-proofing had failed and the mice ruined the plants as well as making themselves very obvious: the row of aviaries was finally converted for the use of small tropical hawks and owls, so that if any mice care to come in now they are more than welcome! In the Reptile House a snake has been used, quite accidentally, to deal with the mice: a small West African house snake—a harmless variety which takes only small rodents—escaped in the laboratory, to be found many months later several times its original size, having preyed on the mice that can never be completely eradicated. In the same house larger geckos are found useful in clearing up cockroaches in cages where it is not wise to use insecticides.

The use of introduced beasts of prey to tackle immigrant pests in Australia and New Zealand has been discussed in the previous chapter. Those responsible had the general idea of biological control in mind, but they had not yet realized how great were the potential dangers of upsetting further an already unbalanced state of affairs. In Australia the fox had absolutely no effect on the millions of rabbits, and it was not very long before it was an even greater menace itself, destroying both farm stock and the rather unsophisticated marsupials.

Among the recent measures tried against the rabbit in Australia is the introduction of a mosquito-borne disease, caused by a virus and absolutely confined to rabbits: this latter point was checked very carefully at the start. The rabbits can also spread the disease by contact, but it is most efficiently spread by the blood-sucking mosquito. A recent article by F. Lyons in the *Geographical Magazine* suggests that this disease, called myxomatosis, is most damaging in river valley areas: in places it is killing up to 90% of the rabbit population, a proportion that is really significant—unless the rabbits manage to develop a strain resistant to the disease. But perhaps this action savours more of biological warfare than biological control.

In New Zealand they tried members of the weasel family chiefly, but the results were equally disastrous, for the unique native birds, including several flightless species, suffered far more than the rabbits. It was partly because of these misguided attempts at redressing the balance of nature that these two countries have now passed such strict laws concerning the importation of animals.

There had been one much earlier experiment along these lines. The Roman colonists in Spain found the native rabbits a pest in their farms: they reported to headquarters and were sent some of the recently domesticated ferrets for the job, but the result does not seem to have been recorded.

Jamaica was the scene of several early efforts at biological control. Black rats had arrived there by ship in the early days and they became a real menace to the sugar-cane plantations, even making it impossible to farm some parts of the island. As long ago as 1762 an ant (*Formica omnivora*) was brought in to prey on the young rats—one of the very few cases where insects have been used to control higher animals; it worked for a bit, but then the effect became less noticeable and the ants themselves became a pest, as their Latin name suggests they well might. Next the Jamaicans tried a giant toad (*Bufo marinus*), with no effect at all; that was in 1844. After discussing it for a long time, the planters finally introduced the Indian mongoose in 1872. At first the rats suffered severely and the damage done by them was greatly reduced, but within ten years it was obvious that a fatal mistake had been made, for the mongoose was an omnivorous predator and attacked the ground-nesting birds and waterfowl as well as the snakes and lizards that were themselves good rat-catchers. This same mongoose is held in considerable esteem in its homeland for its snake-fighting, and a tame Rikki-tikki-tavi is often kept about the house to deal with any cobras that may come along.

It must be admitted that none of these experiments of introducing mammals to control mammals has been successful, and practically all serious work has concerned insects, with a few other invertebrates. In every case nowadays the greatest care is taken to ensure that the introduced animal will feed only on species which man considers noxious, and will not turn its attention to other things. The controlling agents employed are mainly insects which are known to have remarkably fixed habits; so that even when their normal food (the pest) is rare they will not attack other beneficial species. Biological control is not a universal remedy. In fact, it can be used only in a small minority of cases, but where it succeeds it often does so in a most spectacular and satisfactory way. It stands most chance of success where the pest is a specialized foreign invader, and there seem to be no records of victories in any other conditions. As indicated above, such insects or plants—chance introductions in plant, stock, merchandise, etc.—often find an unoccupied niche awaiting them, with no competitors and no enemies; in other words, ideal conditions for them to become large-scale pests.

Some sporadic efforts at biological control had been made in Europe in the first half of the 19th century, and in 1883 a predatory mite sent to the United States attacked the imported cabbage-worm with some success, but the first resounding victory was scored in a warmer climate, about 1890. Very extensive citrus plantations had been developed in California, and insect pests were already becoming a problem. One of these, called the cottony cushion scale, was such a serious pest that it threatened the very existence of the citrus industry, for insecticides had entirley failed to control it. (We have a number of scale insects in Great Britain, especially on orchard and forest trees; they are rather degenerate members of the same insect order as the more familiar aphids or green-fly.) This citrus scale was known to have come from Australia and so a search was made for possible enemies in its homeland. A little ladybird, later to be known as Vedalia, was selected for test, and the first party reached California in 1888. Within eighteen months the citrus orchards were practically clear of cottony cushion scale, and biological control was born. This ladybird has since been introduced into many other citrus-growing parts of the world and has everywhere conquered this scale.

Ladybirds are ideal for insect control; they are generally rather specialized, attacking only one species or a narrow group of species, generally harmful to man's crops, and both adult beetles and larvae join in the attack. Other kinds of ladybirds have been called in, mostly to deal with scale insects and mealy bugs, but insects in several different orders feature in other experiments. Several minute wasps, so small that an adult develops completely inside its prey, have also helped to make the control of citrus mealy bugs more or less complete, and similar wasps have been used in Hawaii to combat the sugar-cane leafhopper (something like the froghopper), the sugar-cane borer (larva of a small beetle), and the fern weevil. Perhaps a weevil feeding on ferns does not sound very serious, but this one, an immigrant from New South Wales, set to work so vigorously that it threatened to interfere with forest cover in forest reserves and cause serious erosion.

It was suggested above that the object is to reduce the damage to below the point where it is economically serious, and not necessarily to exterminate the pest. This has happened, for instance, in the case of the gipsy moth introduced from Europe to the U.S.A., where it became a serious defoliator of trees: a mixed army of parasites and predators was enlisted and introduced which by 1928 were causing a total mortality of over 90%. The same result was obtained with the alfalfa weevil, which reached the U.S.A. by chance in 1907. A tiny parasitic wasp was introduced in 1911: by 1914 it had overtaken the steadily advancing army of weevils and was causing a 90% mortality.

Biological control has been employed elsewhere in dealing with other forestry problems, almost without exception on trees growing out of their natural habitat. One of these experiments concerned a large wood-wasp, Sirex, accidentally introduced into New Zealand, where it did a lot of damage to the imported Oregon pine. This wood wasp, yellow and black and big enough to be mistaken for a hornet in flight, confines its attention to coniferous trees—pine, spruce, larch and silver fir mostly. The adult female deposits her eggs in the solid wood and the larvae spend up to three years eating and growing, in this time making tunnels as long as twelve inches and doing much damage to the timber. In this country it is of no great importance, and its attacks should be taken as a sign that a plantation is in poor condition, but in New Zealand it became an economic pest.

Two parasites, both of them wasps, are native enemies of the wood-wasp, and the larger of them (actually the largest kind of British ichneumon and as much as four inches over all) was selected for trial in New Zealand; its scientific name is *Rhyssa persuasoria*. This wasp has the uncanny ability to spot its prey, the growing wood-wasp grub, deep down in the timber, and to bore down with its ovipositor so that an egg can be placed on the grub. This hole may be over an inch deep, and Dr. R. N. Chrystal, of the Imperial Forestry Institute, Oxford, who was responsible for the research on *Rhyssa*, noted that the drilling of such a hole took less than twenty minutes. He found that *Rhyssa* sometimes drilled down to a section of a tunnel no longer occupied, but never wasted time at a place where no larvae had been. A tiny grub hatches from the egg and attaches itself to the host, which it gradually eats; it takes nearly a year to develop, though it is resting for much of this time, and about May it emerges from its cocoon to cut its way out of the tree and start life as an adult insect. For the control work a large dump of logs known to be full of wood-wasp grubs was made in an area where *Rhyssa* lived; these were soon thoroughly parasitized and in the autumn the logs were sawn up to obtain a good supply of the parasites. These were put in small gelatine capsules and sent to New Zealand, where they soon developed and set to work.

Other examples from forestry of the use of insects to combat insects have shown moderate success, but there have not generally been such overwhelming victories as those scored in the fields of agriculture and fruit-growing. Practically all successful cases so far quoted have concerned specialized insects having a narrow niche and having equally specialized enemies available. An experiment now going on has as its subject a generalized pest, the giant snail, that was dealt with earlier as an invader. Efforts are continually being made to find efficient parasites or predators for this snail which has colonized so many parts of the Far East, and the latest ally to be tried out is a small carnivorous snail from the coastal zone of Kenya, the homeland of the giant snail. Dr. R. T. Abbott, who had the job of enrolling a small army of these snails, described it as a David very able to slay Goliath *Achatina*. He was able to collect and transport several hundreds of these little snails, and just now they are fighting an experimental war on a tiny island near Guam in the Pacific Ocean. If the Davids win they are likely to be tried in force in areas where the giant snail is a major pest.

Man has also enlisted insects to help him control injurious plants. This problem is a very knotty one, for the greatest care must be taken to ensure that the insect cannot get out of hand by adapting itself to feed on cultivated plants or useful wild ones. Elaborate techniques guard against this.

To start with, only very specialized feeders are considered, and before any candidates are imported they are tried out with all plants of economic importance in their homeland by being given the choice of feeding on these plants or starving. If they pass this stiff test they then move to their proposed new hunting ground and are similarly tried out there on all the economic crops; this testing is done in insect-proof laboratories, so that there will be no accidental escapes. If an insect passes this test, too, there is virtually no possibility of its becoming a pest.

The classic example of success in weed control by insects occurred in Australia in the middle 'twenties, when the prickly pear was at last conquered. Several species of cactus, all natives of the New World, had been introduced into Australia from time to time as hedge plants, but the real scourge developed from a single plant of *Opuntia inermis*, brought to New South Wales in 1839. For some time it behaved itself, but about 1870 this prickly pear got right out of hand. By the end of the century it covered 10,000,000 acres; by 1920, 60,000,000, and it was then spreading at the rate of some 10,000,000 acres per annum. As the ground was colonized it became quite useless for grazing.

Meanwhile the search for the right insect had been going on, and several had been tried with a certain amount of success, but it was in 1925 that the real answer was found. This was the tunnelling caterpillar of a tiny moth native to Uruguay and northern Argentina, and very expressively named *Cactoblastis cactorum*. The prescribed tests were carried out, and only one shipment of 2,750 eggs was made. In 1926 about 9,000,000 eggs had been produced in special breeding establishments and distributed. The caterpillars found no competitors at all and no enemies—for great care had been taken to introduce none of their parasites—and they got down to their work of eating out the fleshy stems and letting in all kinds of fungi and bacteria. The result was indeed spectacular, as whole areas of cactus died completely and again became available for farming.

The victory is even commemorated in a poem published in the *Cactus and Succulent Journal* of America, but the caterpillar is such a specialized feeder that the implied fear is actually groundless.

> So abandoning disguises
> Cactoblastis chews away,
> Till another problem rises
> To confront another day:
> When the pear pest in the past is
> Who will blast the cactoblastis?

In California a battle is now being won over a familiar British weed—one of the St. John's Worts—that first appeared there in 1900. It began taking over pastureland at the expense of the grazing plants and was itself poisonous to sheep and cattle. Australians had already been working on the same problem and had located possible allies, all of them beetles; after undergoing a further probation, one leaf-beetle (*Chrysolina gemellata*) was set to work in 1945. In less than three years twenty acres of dense growth was completely destroyed and at the same time enough beetles had been reared to set up colonies over a wide area. The general effect is rather like a slow forest fire with the flowering St. John's Wort being killed by a slowly advancing wave of beetles. The whole job will take several years, but the end of St. John's Wort as a pest in California is probably in sight.

The thorny shrub Lantana is a native of Mexico, but it has been introduced to many tropical and sub-tropical countries as an ornamental hedge. Where conditions have been just right for it, this shrub has got out of hand and has covered large areas. This has been especially true in India and in the Hawaiian Islands, where it was taken about 1850. Early this century, as a result of an expedition to Mexico, a team of insects was enrolled to attack Lantana and eight members of the team established themselves. Each has a slightly different line of attack, but between them they make fairly certain that though a few fertile seeds may still mature on most Lantana plants, the damage to leaves, fruits, flowers and tender branches is so continuous that it is no longer the pest it was.

Work on biological control goes on all the time and most Research Institutes have special sections dealing with it. Perhaps we are unlikely to see many more of the really spectacular victories such as that scored by Vedalia: the strictest control of imported plants aims at seeing that such situations do not develop. But as man understands better the interaction of the thousand-and-one different creatures composing the natural balance that he so rudely disturbs, so he becomes more skilful in getting animals big and small to help him achieve his objects.

Part Three

Animals in the Service of Man

1. PRODUCERS OF FOOD AND CLOTHING

THERE is a big and very assorted group of animals which have been brought into the most important of all relationships with man—they have become domesticated. This relationship is now world-wide and there can be scarcely a human group today which does not keep several kinds: the only exceptions, if there are any, will be found among the primitive, completely nomad tribes, and even then it is a case of having lost such as they once had rather than of never having had any. For these animals one can use the words "domestic" or "domesticated", but it should not be assumed that a domestic animal is, literally, one that lives around the house. Most people would find it easy to name quite a number of these domesticated animals though they might not realize that there are more than forty different kinds of them, varying from insects to mammals, but they would probably have some difficulty in defining one. One fairly old detailed description reads, "Those animals which form part of a household, which are under the domination of a master to whom they give their produce and services, which reproduce themselves in their state of voluntary captivity, and produce young which, like themselves, are attached to the household and become servants of the master." One short modern definition is "a form living in symbiosis with man", symbiosis being a state in which each organism is more or less dependent on the other.

In general, domesticated animals differ considerably from

their wild ancestors, though there are numerous exceptions; in many cases, too, they are derived from animals which are gregarious in the wild, a fact which may well have assisted in the first case of bringing them into this intimate contact with mankind. It is also interesting to note that in a number of cases, notably the horse, cow and sheep, the wild ancestors have become almost or quite extinct. There is distinct difference between tameness and domestication; single animals of a very wide range may be tamed, including many most unlikely species, but whole species are not thereby domesticated. At the same time not by any means all individual domestic animals can be described as tame.

Man alone among the higher animals brings other animals into his direct service by domesticating them. It might perhaps be suggested that birds such as cuckoos that get their incubation done by other birds are in effect employing foster-parents, but this is nothing more or less than parasitism. Among the ants and termites, however, there is something very akin to domestication and pet-keeping. The relationship between ants and green-fly (or aphids) can be observed easily enough in Great Britain; too easily, in fact, for the ants are partly responsible for setting up new colonies of fly on the tender shoot-tips. It is rather like establishing a herd of cows on a fresh pasture, for the ants visit the aphids regularly to drink the drops of sweet liquid which they secrete on request. In tropical countries the ants go much further than this, for they confine their "cows" in special underground pens where they feed on the sap in small rootlets: these pens have entrances passable only to the attendant ants. In many nests of both ants and termites a variety of "guests" can be found, belonging to very many insect orders. Most of these "guests" are never found outside the colonies and this seems a true form of symbiosis, for they are fed by their hosts (or feed uninvited on the hosts' eggs and young) in return for services rendered in the form of tasty or sweet morsels which they secrete. The "cow-keeping" habits of ants may have serious consequences: for instance, the virus causing the deadly swollen shoot disease of cocoa in the Gold Coast is transmitted by a bug which in turn is carried round by some tree ants!

According to the first definition above, a truly domesticated

A painting of a water buffalo ascribed to the 12th century Chinese painter Liu Sung-Nien

British Museum

Ploughing matches are still popular in many parts of the country

animal breeds freely in captivity: but there is also a further small class, typified by the elephant and the falcon, which is truly brought into man's service but does not normally breed in captivity. Animals which serve man, whether in the former or in the latter class, can be put very roughly into three categories: (1) providers of food and clothing; (2) transport animals; (3) pets, companions and allies. Many fall into two or even three of these groups and comparatively few into one only: many animal orders are represented in (1) and (3), but transport animals are obviously drawn only from the mammals.

Domestication began in the fairly early ages of man, and nothing has been recorded to show just how it took place. It seems that there was a fairly widespread human movement and that in some cases, perhaps many, species were being brought into domestication at several different places, sometimes more or less simultaneously, sometimes at wide intervals, so that few domesticated animals had a single origin. This movement was probably at its height between two and four thousand years B.C., and practically all of the important Old World kinds were already established by 1000 B.C.

It seems to be generally agreed that the dog was the first animal to become intimately associated with man, and it was probably the first by many hundreds of years. Among the animals whose chief job was to provide food, and also clothing, the order was ox, goat, sheep and swine, while in the transport animals of the Old World it was apparently camel, ass and then the horse last of all. The Old World was the origin of by far the greater part of the important domestic types: this is partly because the Middle East was the cradle of the early civilizations, but an equally important point is that suitable kinds of wild animals were available. As in so many spheres, opportunity is the great thing: this lack of opportunity largely explains the poverty in domesticated animals among the forest tribes of Africa, for there is virtually nothing suitable for domestication among the local animals and most introduced kinds cannot resist the local diseases. The whole of America has added only three kinds of domestic stock and Australia has made but a single contribution, the budgerigar, which is included among the companions of mankind.

The importance of animal domestication in the development

of the human race could not be exaggerated, but its discussion would be out of place here: it is really a subject for the anthropologist. In the realm of transport, however, the advent of draught animals was about as revolutionary as the coming of the railroad and then the internal combustion engine have been to a much later era: the great human migrations would have been quite impossible without animal transport. The domestication of the cow, sheep and goat, with the development of a pastoral life, made it possible to occupy new areas in the Middle East several millennia ago: in the same way ranching and sheep-farming were the main reasons for opening up great stretches of country in the Americas, Australia and New Zealand.

Europe has some claim to be the origin of the domestic cow, which is, without doubt, the world's most important animal: it is fairly certain that its ancestor was the aurochs, *Bos primigenius*, now extinct. This great beast, standing some six feet in height and very long, had died out undomesticated in prehistoric times in Great Britain, but it survived in the remote parts of Poland, alongside the European bison, until the 17th or 18th century. There is, however, much argument about the cow's geographical origin, and some authorities consider it more probable that it was domesticated in the Middle East and distributed from there, possibly being crossed with local wild forms later. There must always be argument as to how such a formidable beast was subjugated: certainly it would have been impossible to catch and tame adults in their full strength. One novel suggestion is that the domestication of the cow was a sequel to its religious associations. The cow certainly has a long connection with various religions, being held in great reverence by some peoples and sacrificed by others, and often associated with the moon because of the shape of the horns. The disappearance of the moon in an eclipse was considered a great disaster that called for sacrifice, but as an eclipse could not then be foretold it was often difficult to get hold of a wild ox in time: as a result it was decided to enclose a herd for such emergencies and from this came the cow's domestication. We have no possible way of knowing today what actually happened and this suggestion seems as reasonable as any.

The taming of the ox was fundamental in the establishment of an agricultural existence, and it is often considered the greatest achievement of early man, for the ox made two vital contributions: it gave milk and it pulled a plough. Cows were known in the earliest dynasties in Egypt, though by the way they were secured—as shown in ancient carvings—it may well be that they were still half savage then. The oldest known Indian documents refer to milk and butter, and to the sacred character of the cow. Neolithic man brought with him to England a small dark red beast called the Celtic ox, but both man and beast were later driven west and north by the invaders from the Continent, who brought with them the tamed aurochs. At this distance it is very hard to know just what these early cows looked like, but it is possible that the tiny groups of British park cattle are their direct descendants. There are three main types existing today, known as Chillingham, Chartley and Cadzow and, although they differ in various details, they are all white, except for black or red ears and black muzzle and hooves, and with very long straight backs. Each of these herds has been segregated for centuries, and those at Chillingham Castle are known to have been enclosed about 1240. But whether they came from the aurochs or are the descendants of some form of the Celtic ox specially developed for religious purposes we shall probably never know.

At first the ox was valued as much for its draught qualities as for anything else, but although oxen may still be seen at work in Italy and southern France they are now little used except in parts of tropical and southern Africa: the trek ox actually played a great part in opening up South Africa. Today the cow has many different uses, but there are three principal ones—the production of meat, milk and manure: this last may not seem to be very important but it is, in fact, of tremendous value, especially where the cattle receive a good deal of their food from outside the farm. This manure is largely independent of the type of cattle, though obviously it can be collected and then distributed only when the cattle spend a considerable part of their time in stalls. Modern cattle have been bred to produce meat and/or milk in a variety of climates: a few beef breeds are quite useless as milkers, but some are distinctly dual-purpose animals and the milk breeds generally finish up as beef. The black Aberdeen-Angus and the red-and-white Herefords are the outstanding British beef cattle, while the Shorthorns are the best general-purpose group: for milk production Ayrshires and Friesians give the quantity but Jerseys and Guernseys produce fair volumes of high-grade milk very rich in colour and butter-fat. The seven breeds mentioned are by far the most important, but some twenty main pedigree types are kept in Great Britain, the result of centuries of segregation and careful selection.

Cattle have always been a mainstay of human life, and in parts of Africa today there are still tribes which live on little other than milk and blood drawn from their cattle; at the same time their animals are very important as visible wealth. The African cattle, however, are thought to spring largely from another stock—zebu—which itself probably came from the wild banteng of south-east Asia. There has been great argument about the origin of this tremendously important humped cattle, which according to ancient drawings had already reached something like its present form well over 2000 B.C. Some authorities think it came from a long-extinct wild ancestor, but Antonius, of the famous Schönbrunn Zoo at Vienna and a great authority on all wild cattle, is satisfied that it comes chiefly, if not entirely, from the banteng, as does the Bali cow. The gayal is another form of domestic cattle found in parts of India and Burma: it is derived directly from the wild gaur, a fact only confirmed fairly recently by finding that they agree in such unique habits as, for instance, in flinging the whole body sideways when attacking instead of going head first. This observation was made in the Vienna Zoo.

South-east Asia is also the home of the water buffalo, which must be rated the most important domestic animal in the tropics; it is estimated that there are 20,000,000 in India alone. The water buffalo is of ancient origin and is shown in drawings four to five thousand years old. Invaluable for cultivating the rice fields, it is also used for ploughing throughout Asia and even in the Danube Basin; in addition it provides excellent milk and meat. Several types of domesticated buffalo may be found, but the one common in India is indistinguishable from the wild form, with which it breeds if allowed to do so. It is very much a swamp-loving animal and spends most of the day more or less submerged in mud. Buffaloes seem very intelligent, and it is recorded that one in Malaya was taught to take a bamboo bucket to the river, carried on the horn, and bring it back full. Others have been taught the sort of tricks more usually associated with circus horses.

Far to the north, along the Himalayas, the yak was tamed by the Tibetans many centuries ago, and today, little different from the wild form except in coat and colour, it still serves its masters well, giving milk and meat besides carrying them and their goods over the high passes. For this last work it is quite indispensable. Nobody knows when the yak was domesticated: the long, very well-haired yak's tail made the ultra-fashionable fly-switch for the ladies of Rome under Domitian, but that is no proof that the yak was already domesticated.

The sheep is another animal of tremendous importance to the human race, and in some countries it is far more important than the cow. Like cattle, sheep have many uses but three principal ones—the production of meat, wool and manure. The sheep appears on the scene later than the cow and goat, but was domesticated very long ago, well before any history was written. Strangely enough, we do not know with certainty what were its wild ancestors. It seems that there may have been two centres, for it is agreed that the moufflon of southern Europe still found in Corsica and Sardinia comes in to some extent (it certainly hybridizes freely with domestic sheep), as do also some of the West Asiatic wild sheep, but none of these has a true wool. Much research is being done on this problem in connection with the vast sheep industry of Australia, but so far the origin of the wool is untraced; there are two possibilities: (1) that one of the domestic sheep's ancestors is long extinct, and (2) that the woolly coat was a sport or freak which turned up very early and became established as a permanent character.

Some of the primitive sheep have long fat tails: others store their reserves as fat in the rump. The wild ancestors were presumably coloured, as is the moufflon today, but a white coat was soon developed, though black sheep crop up from time to time in white breeds and black faces are a character of such breeds as the Suffolk and the Scottish Blackface. These are but two of the thirty or so varieties of specialized sheep found in Great Britain today: the Soay sheep, seen today in most zoos, indicate the type of stock from which these flocks are probably descended, with an admixture of outside blood. Some of these special forms were more or less established four or five centuries ago. In this country there are three main types: the lowland, the downland and the mountain breeds; each breed has been developed to produce meat and/or certain types of wool in localities varying from marsh to mountain. In New Zealand and Australia this process of selection has gone still further. The wild sheep are all horned, and this feature is still found in a few domestic breeds: these horns are always more or less spirally coiled, in contrast to goats' horns, which are flattened and backwards pointing, not curled. Like the ox, the sheep is primarily a grazer, i.e. it eats grass, though it has learned to take roots and a variety of cultivated produce, in contrast to the coarser-feeding goat, which prefers to browse, i.e. feed on tree leaves.

The domestic goat is principally descended from the wild goat of western Asia, which is still found on some of the Greek islands. In western Europe it is not often of major importance and it is only used locally for milking; but in Mediterranean lands, in the Middle East and in parts of Africa it occupies a much bigger place in the local economy, and is valued for its hides, milk and meat, though the meat from an adult male is hardly edible. Special breeds of goats yield the famous Angora and Cashmere wool, but only in suitable climatic conditions. Although there are no true wild goats in Great Britain, a number of domestic animals have escaped into the hills to found semi-wild herds and in western Scotland there were "wild" goats nearly 200 years ago. During World War II the goats on some of the Irish hills were shot to feed the lions in the Dublin Zoo. Goats are gregarious animals and are most intelligent and brave: Col. Jim Corbett, slayer of many man-eating tigers, has a remarkable film taken in North India of a billy goat fighting off successfully a leopard that tried to kill it. At the same time the goat, with its sure-footed ability to climb anywhere and everywhere and eat anything and everything, can be the forester's worst enemy, and it is often responsible for serious erosion, especially in Mediterranean countries. Goat breeding has not generally been taken very seriously in this country, but there is now some effort to develop definite types, as has been done in Switzerland.

The pig is the last of the really large-scale producers of food. In recent years it has greatly increased in importance as it provides one of the most rapid and economical ways of turning vegetable food—including all sorts of waste and swill—into meat and animal fat; it can put on a steady pound per day and it is killed at the age, and therefore weight, appropriate for the market—that is, according to whether it is wanted for sale as pork, bacon or ham. Pennant, the British zoologist, suggested nearly two centuries ago that the pig was not inaptly compared to a miser, who is useless and rapacious in his life but on his death becomes of public use! The pig, of course, produces its quota of ever-useful manure during its lifetime, but the only real purpose of its existence—as far as man is concerned—is to become fat and meat. The pig's ancestors—the European wild boar, the wild pig of south-east Asia and the Chinese pig—are omnivorous scavengers with a snout especially suitable for surface digging; these characteristics remain in the domesticated form, and (in view of the danger of carrying disease) they are the real reason for pork being proscribed in Levitical law as food for the Israelites as well as being forbidden to the other Mediterranean peoples. Domestication took place in early days, perhaps in the third millennium B.C., but after the ox, goat and sheep had been enlisted by man. Outside these Mediterranean areas the pig has always been a valuable source of food, and it had the great merit of fending for itself for much of the time. Pannage, the right to let pigs run loose on common land in search of acorns and other food, goes back many centuries: the word "pannage" itself is over 500 years old and it was derived then from an earlier Old English stem.

The gigantic sows having litters of about twenty are very different from the long-headed, razor-backed hogs hunting a living that our forebears kept. Modern selective breeding has developed about a dozen different types suited to various conditions and needs, but the pig's job is just to produce meat and fat and it has therefore been much less modified than the dual-purpose sheep. Perhaps the strangest thing is that the domesticated pig has completely lost the lateral stripes—really a series of dashes—that one finds in the young of almost every known wild pig, certainly in all those included in the ancestry of the domestic pig. In some parts of the world, particularly in the south-west Pacific, pigs are often allowed to run loose around the villages, seldom if ever being penned, and they tend to revert to something like the original wild type: some indeed take to the woods and become completely feral.

After dealing with several animals whose origins are obscure, or to say the least rather mixed, it is a relief to come to the reindeer, for there is no possible doubt about its ancestry. It is, in fact, almost indistinguishable from the wild reindeer, or caribou. In many ways the reindeer is unique: it is the only deer to be domesticated, it is the only deer in which both male and female have antlers, and it ranges much farther north than any other deer, living within and just outside the Arctic Circle. It is a specialized animal in many ways, and for this reason it is of use only within its actual homeland: in this it resembles other specialized beasts such as the llama of the Andes and the yak of the Himalayas, but differs from the great majority of domesticated animals. There are several slightly different forms in the far northern parts of both Old and New Worlds, the wild form in America generally being called caribou. Several of these races have been domesticated by nomads in northern Europe and Asia, but none in America, though domestic reindeer have been introduced there, with little success so far, in recent years.

It has been assumed that the reindeer was tamed a very long time ago, and this may well be so, though there is no real proof of it. Chinese texts of some five centuries B.C. refer to deer being kept by the barbarians of the north, but while these probably referred to the reindeer, there is some uncertainty, nor is it known whether they were used for carrying baggage, drawing sleighs or, as in Siberia, for riding. Reindeer are specialized feeders, at some seasons largely depending on reindeer moss (*Cladonia rangifera*), a lichen found throughout the tundras; they can find it under some three feet of snow and dig down for it. They also browse a lot. They live in large herds —always in the open—watched over by their owners, and it is estimated that these herds total some 1,500,000 head. They are valuable also for their milk and meat, and the introduction of reindeer into parts of Scotland has been considered more

A team of pack donkeys in North Ashanti, Gold Coast

G. S. Cansdale

From time to time zebras are broken in as draught animals

Paul Popper

Elephants were used in war many centuries ago

Anne Cook

than once. A project is being planned again just now and Swedish experts are co-operating: it would, in fact, be a reintroduction, for reindeer were lost to Scotland only some 700 years ago, largely due to progressive climatic changes. The areas in Scotland at all suitable for reindeer are very small and it is impossible to hope for the establishment of more than a few small herds.

As with the reindeer, there is no doubt at all about the origin of the tame rabbit, for it comes from the European wild rabbit; although it is now so widely distributed as a wild animal, it is generally thought that it was at one time confined to Spain. It reached France in the Middle Ages, probably introduced as a sporting animal, and first appeared in Great Britain in the 12th century. It was certainly not known in ancient Greece and Rome, but it is generally agreed that it was domesticated by the Romans early in the present era. The Second Punic War gave the Romans control of Spain about 200 B.C.

It is possible that an Asiatic rabbit has contributed to the domestic rabbit stock, but it is quite definite that the hare has had no part in it, in spite of the names of such varieties as the Belgian hare: the two animals are very different in many ways and cannot interbreed. The wild rabbit has several natural colour varieties and it is obviously a species that produces many sports, or mutants. This fact, and the early maturity (about six months), the large litters (sometimes over a dozen) and the fairly short gestation period (one month) have all helped to bring about the almost fantastic assortment of pedigree rabbits now kept in Great Britain and other countries. The development of fancy breeds is fairly recent, and one popular type, the Rex, is derived entirely from a mutant which occurred in France in 1927: this kind has a close short fur rather like a mole's. Further mutations have already arisen from this one and pure strains of almost every colour have been fixed.

The first use for the tame rabbit was presumably for food; in many other countries this has always been and still remains its principal use, but the wild rabbit has been so plentiful and, until recently, so cheap in Great Britain that the tame rabbit was generally kept either for fur production or as a pet, principally the latter. The largest rabbits of all now weigh over

20 lb., against about 3 lb. for a wild rabbit. Among the special fur-bearing rabbits the Blue Beveran is still popular, but vast numbers of natural skins are treated to imitate a great range of expensive natural furs. One wool-bearing rabbit has been developed—the Angora—in which the long coat is clipped every few months and used for the finest of knitting.

In very recent years the rabbit has become valuable as a laboratory animal: it features in one of the standard pregnancy tests and its broad flanks are ideal for standardizing such products as the cow-pox lymph used in vaccination.

With the rabbit, fur-bearing is only one of the reasons for which it is kept by man, but there is also a whole range of animals kept solely for their furs. The growth of this comparatively new business of fur-farming is of particular interest, for in it one can see several distinct stages of domestication, from the pure wild form simply confined, on the one hand, to a wide range of colour varieties, very different from the wild animal and fully fixed as breeding forms, on the other. The coypu, whose undercoat yields the thick, more or less waterproof fur generally known as "nutria", is the best example of the first group; an inhabitant of the waterways and swamps of South America, it has shown itself able to settle down and breed fairly freely in suitable habitats in the northern hemisphere, including the British Isles. It is notable in having a gestation period of no less than 140 days, an almost fantastic length of time for a rodent, especially when compared with only thirty days for a hare, of about the same size. The coypu is a regular part of the stock-in-trade of the small travelling showman who shows it as "the biggest rat in the world", and may even suggest that he caught it personally in the London sewers! No colour-mutants seem to be known for the coypu, but in the somewhat similar musk rat from North America a black form is more valuable than the natural brown. This musk rat, rather like an oversized water vole, was first imported into Great Britain to found fur-farms, but it settled down so well and established itself so firmly in several valleys, notably the Severn and the Arun, that it threatened to do serious damage to the banks; it was therefore reclassified as an undesirable alien and a special law now forbids its importation and aims at its extermination. It was introduced into Central Europe early this century, but it is a great traveller and it has since colonized a considerable area of Germany and Czechoslovakia. It was taken to Finland after the First World War, but there too it has become a nuisance, again emphasizing the great potential danger of introducing animals to new lands. The trade name for the musk rat's fur is "musquash".

The Canadian beaver, famous as it is for its fur, has no real claim to a place here: it is carefully protected, areas are restocked from time to time and trappers no doubt regard colonies as their own private flock or herds, but they remain truly wild.

At the other end of this group of fur-bearers are the silver fox and mink, scarcely different in form from their still wild North American cousins, but now showing a number of beautiful colours some of which, especially in the mink, fetch almost unbelievable prices. These colour-forms have been obtained by selective breeding from mutants, a rather simpler task than in many breeding experiments, for the only real factor to be considered is the coat colour. A recent estimate of the output of silver fox pelts from North American "farms" was 200,000 per annum. The Arctic, or blue, fox is also a valuable animal in the fur trade, but it has so far proved impossible to keep satisfactorily in confinement and it is raised on small islands along the coast of Alaska. It is allowed to run free, though given a certain amount of feed, and then trapped. The thick whitish winter pelt is the valuable form that is taken.

A new recruit to the fur-bearers is the chinchilla, a most exquisite burrowing rodent from the slopes of the Andes, whose hair is the softest and finest animal fibre known to commerce; this hair was being used as wool before the end of the 16th century and was made into covers and blankets, but the whole skins are now used for making perhaps the most expensive of all fur coats. In its homeland it has been hunted almost to extinction, but it appears to be safely established on fur-farms in the southern States and is thus well on its way to becoming domestic. It is a stocky little creature up to about a foot long, with a short, rather stiff tail, and to preserve its wonderful coat it is generally kept in a special cage so designed that the fur cannot be soiled. In captivity it seems to approach the tabloid-eater so long promised by advanced scientists, for some breeders feed it entirely on small dry pellets of a specially compounded vegetarian mixture.

The ostrich certainly deserves a place among the domesticated animals and it fits in as well here as anywhere, for it approaches as near to fur-bearing as a bird can. It may be a little difficult now to appreciate the volume of the trade in ostrich feathers a few decades ago, when a single lot in one of the twice-weekly auctions at Port Elizabeth would be knocked down for as much as £25,000, and when London wholesalers handled ten to fifteen tons of feathers a month. The fashion for ostrich feathers goes right back to the Pharaohs; up to about 1870 feathers were obtained by hunting, but a Boer got the idea of starting farms for the world's largest bird. Some fifteen years before this the French had succeeded in breeding some captive ostriches in Algeria, but this was not followed up, and the first commercial development was at the Cape. Very soon these farms were springing up all over South Africa and also, in spite of a vigorous export ban, in California, Australia, and southern Europe. By about 1920, however, fashions had changed and most farms had to be abandoned, though a few ostriches are still to be found living in captivity and serving man. There is just a possibility that the ostrich was kept in semi-captivity in early Egyptian times. A panel at Thebes shows huntsmen returning from the chase with a live ostrich, as well as feathers and eggs, while there is a picture of the period 1580–1100 B.C. in which an Egyptian is leading an adult ostrich. When the Greeks first caught sight of an ostrich they exclaimed, "Great sparrow—struthio!"—from which the exclamation "struth!" may well be derived.

We now turn to the very important group of six domestic birds which belong to the farmyard and are generally classed as poultry: they are kept for two purposes only—to produce eggs and meat. Three of these are waterfowl, though the domestic forms may see little water, and three belong to the large family of gamebirds: the common fowl is one of these latter, and with its many different breeds it stands out in importance, for its world output of both meat and eggs must far exceed all the others combined. It enters the human scene quite late, not being found at all in Neolithic remains and having no mention in the Old Testament. It appears towards the end of ancient Egyptian civilization and it was assumed to have been introduced there by the Greeks, but it is known that Assyria paid tribute to Egypt, in the form of hens, as early as the reign of Tethmosis III (about 1500 B.C.). Homer, writing in the 9th century B.C., knew nothing of it, though of course he spoke of the domestic goose, but poultry-keeping was quite common among the Greeks in the 5th century; by the 4th century it had spread far to the west and various types had developed, including a special breed on the Island of Rhodes. The Romans took up poultry-keeping seriously and before the beginning of our era they had half a dozen distinct forms and had invented artificial incubation: in fact the methods of running a poultry farm remained much after the Roman pattern until they were revolutionized during the present century. The quality, however, was far below what we expect today: the Romans had increased egg-laying from the natural level of 30 or 40 in the spring only to a maximum of 60, but today a good hen lays over 200 eggs throughout the year. In addition eggs are now much bigger.

In Greek and Roman times cocks were bred specially for fighting, and the ordinary domestic cock also acted as a timepiece; day began with the first crow, and the third cock-crow of which Our Lord spoke to St. Peter represented a more or less definite hour of the morning. Many centuries later carefully selected cocks still travelled with each caravan in Central Asia to ensure an early start: cheap imported alarm clocks have presumably taken over this duty now. In some parts the rooster is still regarded as the watch-dog of the poultry flock: in Ashanti, for instance, there is a very well-known proverb that "When the cock gets drunk he forgets all about the hawk." It is fairly generally agreed that the red jungle-fowl of India and Burma, little bigger than some bantams, is the wild ancestor of the great range of specialized modern fowls, but some authorities consider that one or two jungle-fowl species from other parts of the East may also have contributed. Selective breeding has produced not only a range of table birds and layers but, among the latter, sex-linked characters which allow the hens to be picked out as day-old chicks.

The modern demand for eggs and poultry meat far exceeds the existing supply and poultry-keeping today is developing into a largely mechanized way of turning assorted foodstuffs, mostly vegetables, into eggs and meat. In 1951 the representatives of some forty countries got together at Paris as the World's Poultry Science Association to discuss even more efficient methods. Poultry-keeping has gone a long way since the first jungle-fowl, little bigger than a bantam, was persuaded to lay its eggs in captivity.

The guinea-fowl is the only African contribution to the poultry; it is, in fact, one of the very few African contributions to the whole range of domesticated animals. Guinea-fowls were taken to Greece about the 5th century B.C. and became fairly common: they were used, amongst other things, as sacrifices by the poorer people. In spite of over 2,000 years of domestication, the guinea-fowl has remained virtually unchanged and it is hardly distinguishable from the wild helmet guinea-fowl of the open woodlands of Central Africa; the only variation commonly seen is some degree of white in the plumage. On the average, the wild bird today is heavier and makes better eating than the domestic form, but the latter fully retains its ability to fly, in contrast to most other types of poultry.

Guinea-fowl somehow became associated with Meleager, a Greek hero famous for killing a wild boar sent to ravage the country, and they were kept in his shrine in Aetalia: this may, indeed, be the point from which they were first distributed. In one Greek fable the sisters of Meleager were changed into guinea-fowl and the plumage is said to show the tears they shed until they died; this association is commemorated in the scientific name *Numida meleagris*. The German name Perlhuhn (pearl-hen) is derived from the handsome spotted feathers. The Romans seem to have been very interested in guinea-fowl and kept at least two species, one of which has now been lost to domestication. It is possible that the common species also became lost and that it was domesticated anew in the Middle Ages. It has never become very plentiful in Great Britain, where it is generally found around farmsteads in a semi-domesticated state, living very much its own life and laying its eggs where it pleases. It easily reverts to the wild state and has done this in various tropical and semi-tropical islands where it had been introduced, becoming a useful sporting bird.

The turkey is the only North American animal to be completely domesticated—none of the fur-bearers qualifies fully yet. Wild turkeys are found from Canada right down into Mexico and Central America, and it is the Mexican form, very little changed by centuries of domestication, that forms the traditional dish on Christmas Day for the British and on Thanksgiving Day for the Americans. The turkey had been domesticated by the Maya peoples long before the first Europeans reached America, and the Spanish found it very common in Mexico. It reached Europe in the 16th century and it is a well-known tradition that turkeys were first eaten in Europe at the wedding feast of Charles IX of France in 1570. This, however, is incorrect: it may have been brought over as early as 1523, and it was certainly appearing on the table between 1550 and 1560, though at first it was compared unfavourably with the more familiar peacock.

The goose is by far the most ancient of the farmyard birds, though it is impossible to say just when it became fully domesticated. Geese appear on inscriptions and vases of the early civilizations, and they are mentioned in the ancient Chinese and Indian records, but it is not certain whether these birds were captive or wild, rather than domesticated. It is fairly certain that tame geese were kept in Egypt, where they were probably considered sacred, and the first mention of their being eaten is in Homer, some 900 B.C. It is rather strange that they are not mentioned in the Biblical record. The Greeks kept a lot of geese and their literature is full of references to them: Homer, in the *Odyssey*, speaks of eagles stealing domestic geese. The Romans also kept them, both for eating and as sacred birds on the Capitol, where they served as useful watch-dogs. The Romans, it seems, knew all the tricks of fattening geese and they appreciated *foie gras*.

Most tame geese are descended from the wild grey lag goose, but it may well be that the other grey geese—whitefront, bean, etc.—have also contributed. White plumage is very common in tame geese and the powers of flight are much reduced, but otherwise they are much like their wild forebears. The rather distinct Chinese goose was domesticated in the Far East from the Chinese wild goose, sometimes called the swan goose. The very different Egyptian goose has also been domesticated, but more as an ornamental bird. Geese are normally kept only for their meat, but they have one unique virtue—they are grazers, and for much of the year are content with nothing but grass.

The duck came into domestication considerably later than the goose, but it is not clear why this was so because ducks were hunted regularly in early times. The Greeks kept ducks though they were probably not domesticated; the duck was associated with Aphrodite, and it is from this that the word "duckie" is derived as a term of endearment. It seems likely that the duck was more or less established at the very beginning of our era and the eastern and central Mediterranean was the centre from which Europe and Africa got it. Away in the Far East the Chinese had also domesticated the duck, perhaps about the same time, though there is little evidence to date it. China, with its rivers and canals, is ideal for duck-keeping, and this has been developed far beyond anything known in the West. Whole ships are fitted out to house the ducks, with pens all round the deck: the ducks seek their own living for the greater part and the boat moves on regularly to new pastures. These ducks are well disciplined and learn to return promptly when the evening is sounded: the first ones home get some rice, but the last one gets beaten! In China considerable emphasis is placed on egg production, but in most parts of the world ducks are valued as table birds rather than as egg-layers.

The common mallard, found in slightly different forms over much of the northern hemisphere, is the stock from which all domestic ducks have come; among the very few different breeds that have been developed there is a strong white factor. As with fowls and geese, the domestic ducks average a lot heavier than the wild stock and they have largely lost the powers of flight. A few domestic or semi-domestic ducks are trained as a sort of fifth column for use in duck decoys: they are the birds who lead the wild ducks up into the decoy-pipe, but retreat safely themselves as soon as the keeper appears behind them.

Over in Peru another very different duck was domesticated many centuries ago; its commonly accepted name, muscovy duck, is confusing and all sorts of countries have been quoted as its home, but it quite certainly comes from South America. The word "muscovy" refers not to Moscow but to the musky

British Museum

A relief in the British Museum showing camels used in the 7th century B.C. by the Assyrian king Assurbani-pal in his North Arabian campaign

Peruvian pottery many centuries old shows a bridled llama's head

British Museum

Elephants working at a teak mill in Burma

Paul Popper

odour of the flesh at some seasons. The first Europeans to reach America found it well established in Central America. Later explorers found it in many other countries, and in Paraguay the Indians allowed it into their huts to keep down the crickets. The muscovy duck is well suited to life in warm countries and is likely to be useful in areas where other poultry fail.

Several insects must be listed among the domestic providers of goods, but of these only the silk moth can be described as completely domesticated and entirely dependent on man for its very existence. Silk is produced by its caterpillar as it spins its cocoon, and if the silk is to be taken in a condition fit for use the pupa (into which the caterpillar, or larva, has turned) must be killed. Only enough pupae are left, therefore, to hatch and become the moths which will lay the next generation of eggs. The silkworm was cultivated in the early days of ancient China, and it is known that Si-ling, wife of the famous Emperor Huang-ti (2640 B.C.) encouraged the industry. For several thousands of years silk was a product of the utmost importance to China's economy and her most valuable export. Silk production remained a monopoly of the East until some silkworm eggs were successfully smuggled to Constantinople early in the 6th century A.D.; this was the source of all the European silkworms. Spain became a great silk-producing country and the work was mostly in the hands of the Moors; when the last of them left Granada, the centre of the industry, in the 18th century, the silk industry there collapsed. However, it was not from Spain but from Sicily that silk production was distributed through Europe, and the Saracens were the people chiefly responsible. Some silkworms seem to have reached England during the reign of Henry VI; subsequently James I and then George I tried to establish silk production commercially, but nothing has come of it, largely because Great Britain seems to be too cold and damp. The invention of rayon, nylon and other modern synthetic fabrics has had a profound effect on the silk trade, but pure silk remains in great demand for many purposes.

Bee-keeping probably goes back even farther than the silk industry, for a bee is shown as a hieroglyphic in an Egyptian sarcophagus of the year 3633 B.C., but even though it does more or less what is required of it, the bee remains a rather independent creature and few people would claim that by any standard it is completely domesticated. The bee is mentioned in the records of all the ancient civilizations—Virgil and Cicero being loud in its praise—and it is likely that each used its own local bee.

For thousands of years honey was the only source of sweetness, but it also had many other uses. Mead has been brewed from honey since ancient times. Its high sugar content made honey valuable as an embalming material and in its first mention in the Old Testament, in Jacob's instructions to his sons to take presents when they went to Egypt to buy corn, it is in fact grouped with balm, spices and myrrh. Honey also found a use in religious ceremonies in places as far apart as Greece and Peru.

Several different types of bees are recognized today, including the Italian and the Caucasian, but they are apt to get all mixed up. Introduced by man to new lands, bees soon reverted to the wild state and set up their own colonies. In recent years the honey industry has been tremendously developed in Australia, New Zealand and North America, where some beekeepers may have as many as 3,000 colonies and the annual production is of the order of 100,000 tons. In Australia some apiarists mount batteries of colonies in large caravan trailers: by careful planning they can ring the changes in the climate and keep the bees working all the year round. Today beeswax is a valuable by-product, but we often tend to forget that in Great Britain the bees are perhaps as valuable for their work in pollination as for their honey production.

2. BEASTS OF BURDEN

WE now come to a fairly small group of large animals used today primarily as draught and carrying animals, though each of them also has sundry other uses. Of these the horse is by far the most important, though by no means the first to be domesticated. It appeared comparatively late and was probably first domesticated in the third millennium B.C. in southeast Asia, from which it spread quite rapidly over much of the Old World. The first historical record is on a Babylonian tablet of the period of Hammurabi, about 2100 B.C., and the horse is referred to as "the ass from the East". It is absent from Egyptian monuments until about the 18th century B.C., and the oldest European representation of a domestic horse, in southern Sweden, is only a little later. Horses are also depicted in some of the cave paintings in southern France dated thousands of years before this, but it is not seriously suggested that these horses were other than wild, like other beasts shown in the same caves. The ancestors of the domestic horse are not known with complete certainty, but it is more or less agreed that domestication took place independently in three areas—western Europe (western Europe wild horse), south-western Asia (tarpan) and Mongolia (Przewalski's horse). The first of these wild types seems to have existed in Spain until Roman times and it is the main ancestor of the heavy draught horse. The tarpan lasted on the steppes of southern Russia until the second half of last century, the last being shot about 1880. The Mongolian wild horse also seems doomed to extinction, and it may even now be extinct in the wild state: a very few animals are found in Zoos and private collections, but those in British collections have ceased to breed. The only herd with even a remote chance of establishing itself is at the Munich Zoo.

In a recent American book on horses it is suggested that domesticating horses must have been easy; the author goes on to say, "All that was necessary was to catch a few wild colts, raise them as tame animals and in due course let them breed among themselves." Put in that way it may sound delightfully simple, but wild horses in captivity are generally quite unamenable and far worse to handle than most zebras. Taming the first horses was therefore likely to be a tough and long job. The ancient Chinese records suggest that with the eastern Asiatic people, at any rate, the horse was first tamed to provide a food supply, together with the other five of their six basic domesticated animals—namely the cow, chicken, pig, dog and sheep.

The Shepherd Kings of the Hyksos Dynasties were probably in power when the horse came to Egypt; though it is possible that being of western Asiatic origin they introduced it, for it had reached Mesopotamia rather before 2100 B.C. The Egyptians thus already had the horse when the children of Israel moved down to Egypt, and it is a little strange that the nomads did not take any with them at the Exodus. It was not until David's reign that horses were brought to Palestine and it remained for his son Solomon to make them popular. It is not clear where or why the prohibition arose, unless it is simply because the horse does not chew the cud, but the horse is not eaten by the Hebrews or by the Moslems; in many parts of the world both milk and meat are used, and the meat shortage in recent years has brought the horse into the human meat market in Great Britain once again.

When the Romans came to Britain they found a native breed already here with which their heavier horses became mixed. It is sometimes claimed that the Exmoor pony represents this original stock, and this may well be true, though it is generally considered that other blood has been introduced, either purposely or accidentally, from time to time. William the Conqueror's horse was of Spanish blood, while his men rode on chargers of various types; later, the returning Crusaders brought further new blood, including Arab, and, still later, Charles II sent his Master of the Horse abroad to get new stock for the royal stud. Centuries of selective breeding from this mixed stock, compounded of the best available from many countries, has produced the great variety and high quality seen in Britain today, from the Shetland pony of eight or nine hands (three feet) to the Shire horse standing over seventeen hands and weighing a ton.

The past two decades have seen a drastic reduction in the number of draught horses both on the streets and on the land, the inevitable results of mechanization; but, even so, the horse continues to be used for a range of draught and saddle purposes —for work, pleasure and ceremony. What possible substitute is there for the horses which draw the Irish coach and provide the Sovereign's escort on State occasions?

Apart from the four main breeds of heavy draught horses, namely the Shire, Clydesdale, Suffolk Punch and Percheron, there are few breeds as definite as those typical of the cows. Keen racing men would perhaps insist that the Thoroughbred, the name usually given to the horses whose names appear in the stud-book, is a definite variety; but they would have to agree that these horses are bred with one object only—speed; there can thus be considerable variation in both colour and build. The Hackney horses were developed for trotting in harness, but the demand for this type has almost disappeared, and the breed will therefore soon be lost. Hunters are well-bred horses, often rather heavier than the straight racing thoroughbred and thus able to carry a man over rough going and jumps. The word "pony" can mean almost anything, and it is commonly applied to a whole series of small horses from the tiniest Shetland to the polo pony standing nearly five feet; but some serious horse-breeders insist that they recognize no less than eight distinct types—namely Shetland, Highland, Dales, Fell, Welsh, Dartmoor, Exmoor and New Forest.

Introduced to suitable new lands the horse very soon reverts to the wild state, and in many parts of the New World today there are herds of semi-feral horses from which stock is drawn as and when required and broken in for use. Given long enough time, natural selection will sort out a breed well suited to its new homeland. This has happened on the Pampas of Argentina, where some of the fine saddle-horses brought over by the Spanish either escaped or were released in the rich grasslands over 400 years ago. Within a few years there were large herds of "wild" horses, and these became known as the Criollo. They were taken up by the local Indians, to whom they brought vast new possibilities of movement in a rough country. About a century ago the craze for bringing in new blood and the policy of ruthless extermination to make way for cattle threatened to wipe out the Criollo, but action was taken in time to preserve this breed of great local importance.

Although they were long thought to have an Asiatic origin, there is now little doubt that donkeys are a domestic form of the Nubian wild ass, and that domestication took place in Neolithic times in north-east Africa. It is just possible that there has been some subsequent crossing with one of the Asiatic wild asses, but this is unlikely. The donkey appears to have entered Europe by way of Asia Minor, thanks largely to the activities of the Hebrews, who used this animal extensively. It did not reach England, however, until the 9th or 10th century. Coming into the service of man well before the horse, the donkey proved of tremendous value for transport purposes in the eastern Mediterranean area, but it did not increase significantly the speed at which a journey could be covered. The families' loads, and perhaps the wives and children, were carried by the donkeys, but the master seldom, if ever, rode. Today the sense of chivalry has rather declined. It is suggested that the ass was domesticated purely as a beast of burden; however that may be, it is a fact that the donkey has virtually no use except as a working animal.

In many spheres, especially where speed is required, the horse has largely replaced the humble ass, but it still holds its own in parts of North and Central Africa, especially for pack purposes. The donkey is rather unnecessarily maligned; it may be slow, but it is more sure-footed than the horse and less difficult to feed, and it certainly has not deserved the rather miserable slavery that is too often its fate in man's hands. Like the horse, the donkey is able to establish itself if released in suitable country. As long ago as 1591 it was recorded that feral donkeys had reached such numbers on one of the Canary Islands that they had to be exterminated. More recent reports speak of them on some of the West Indian islands and in the Galapagos group. Donkeys average a smaller size than horses, ranging from under three feet at the withers to just over five feet, but comparatively little selective work has been done in breeding the poor man's beast of burden. It is interesting that the very small breeds of donkeys have been developed on islands, especially on Sardinia, in the same way that some of the miniature ponies are also associated with islands. Some of the largest donkeys are referred to as Persian; it is possible that this breed was developed in Persia long after domestication, but this cannot have been its ultimate origin. Colour varies much less than in horses; the body is generally grey, grey-brown or brown, with belly and legs rather lighter but not in any contrasting colour. Most, though not quite all, have the typical black stripe across the shoulders. A white strain is known, also one that is nearly black. Piebald or skewbald markings are very rare, but the London Zoo has a fine skewbald stallion which came from the Sudan in 1949.

The horse and donkey are closely related, and the mule is the result of crossing a jackass and a mare: the hybrid combines the size and strength of the horse with the patience, sure-footedness and endurance of the donkey and is a most useful animal. The production of mules goes back to prehistoric times and it must have been tried very soon after the horse was domesticated. The Greeks and Romans used them extensively, for they do best in hot dry countries. Mules have been described rather cynically as "without pride of ancestry or hope of posterity", for they are completely sterile in both sexes. Theophrastos wrote of fertile mules and other similar claims have been made from time to time, but they are not generally accepted.

The word "mule" has a varied and interesting history. Coming from the Latin through the Old English, it was at first applied just to the hybrid animal, but by 1470 it was already being used for a stupid or obstinate person; this is a little unfair, for the mule's obstinacy is generally the result of its treatment. About a century later Ben Jonson described as a mule a person who was neither one thing nor the other. By the 18th century mule was the general name for animal or plant hybrids, and today it is particularly applied to crosses between canaries and other finches. Towards the end of last century, coin collectors were using the word for coins in which the two faces, for some reason or other, did not correspond. The mule deer, on the other hand, is so called merely because of its ears.

One of the spinning jennies is also known as a mule, but the derivation here is not at all clear.

The camel is very much of a zoological problem. Camels are shown on drawings in Upper Egypt which are dated around 3000 B.C.; that is, about three or four hundred years after the outburst of civilization under Menes and a little before the first pyramids were built. Abraham comes on the scene about a thousand years later and he had camel transport as well as asses and sheep; in fact, camels seem to have formed the main transport for the inhabitants of the Near East, including all the Israelite patriarchs. However, there is no reliable evidence as to where and when the camel was domesticated. It is not even known whether or not the two types known today started off as one and the same animal: these are the one-humped or Arabian camel, of which the dromedary is a "thoroughbred" form used for fast riding, and the two-humped or Bactrian camel. They look different enough and they are suited to completely different environments, but in the skeleton there is little if anything to distinguish them. There is, in fact, less difference between the two kinds than, for instance, between a Shire horse and a tiny Shetland pony. It is known that the Arabian type has been in North Africa and the Near East for a very long time and that the Bactrian camel is at home in the much colder areas of Central Asia, but that is as far as we can go. Some authorities are satisfied that all the camels belong to one species only.

The one-humped camel has certainly never been known as a wild animal in historic times, and there is great doubt whether the few wild herds of two-humped camels found in the region of the Gobi Desert today are truly wild: it seems more likely that they are descendants of escaped animals or even of camels deliberately liberated, for until fairly recent times it was customary for the Tibetans to release camels in connection with certain religious ceremonies.

The Arabian camel continues to be the beast of burden of the North African and Near Eastern deserts, and it has also been introduced successfully into places as far afield as northern Australia; it is able to work in arid areas quite impossible for any other beast of burden, and its only real competitor is a specialized motor column such as that developed by the Long

Martin Hürlimann

Cavalry on the march: a carving in an Indo-Chinese temple built in the 12th century A.D.

Laplanders find the reindeer invaluable for hauling their sledges

Picture Post Library

Paul Popper
A team of husky dogs at work in the North-west Territory, Canada

A shepherd and his dog bringing in a flock of ewes
Picture Post Library

Range Desert Group in World War II. On ordinary journeys a camel can carry up to four cwt., but for such areas as the Sinai Desert only about half that would be allowed. A good riding-camel can cover 60 or 70 miles a day if well fed.

Much has been written about the temperament of the Arabian camel and, it seems, there is much that is better left unwritten; but Kipling's soldier in *Barrack Room Ballads* has given perhaps the best description possible:

> The 'orse 'e knows above a bit, the bullock's but a fool,
> The elephant's a gentleman, the battery-mule's a mule;
> But the commissariat cam-u-el, when all is said and done,
> 'E's a devil an' a ostrich an' a orphan-child in one.

The Bactrian camel lives and works in the dry cold zones of the Central Asian plateaux, and it still provides transport for the caravans crossing the Gobi Desert, if caravans still manage to exist there. But camels are not only useful as working animals: they have yielded milk, meat and hair from very early times and may well have been domesticated to provide these commodities. A camel may stay in milk for up to two years, giving something like two gallons a day at first but dropping to under half a gallon. The hair has many uses and one animal will yield from 4 to 6 lb. Camel dung is a useful fuel in the deserts, and it is even customary in some areas to hang a small basket in a suitable position to collect this as the caravan moves along.

The camel's hump is a storage organ, but contrary to common belief it is used for storing food and not water. It is now known that camels can store water in small vessels all around the stomach (the effect is rather as if the whole inner surface is smocked), though it is not clear just how it works. The comparatively few variations in form seen in camels are apparently confined to colour and general build. The Bactrian is a dark brown; in summer it is rather paler, with shorter hair; on the average the Arabian camel is paler than the other, but true albinos occur and others that are almost black. As regards build, both kinds have heavier and lighter types suited more especially for load-carrying and riding respectively.

The geological record shows that at one time the camel's forefathers were found in many parts of the world and they had lived in America until the Pliocene period, which began about 16,000,000 years ago, but the only surviving relatives of the camels are now found in the highlands of South America; these llamas are placed in the same family as the true camels of the Old World and they might perhaps be described as their distant cousins. They are all much smaller and they have no hump; like the true camels they live in difficult conditions, in their case the slopes of the high Andes above 12,000 feet.

Four kinds of llamas are known—two wild forms with two domesticated forms derived from them and still very like them. The pairs are the wild huanaco and the domesticated llama, and the wild vicuña and the domesticated alpaca. The camel family is thus unique in that all of its existing species have been domesticated. It is impossible to suggest when these two forms were domesticated by the Peruvians, but they were well established as domestic animals when the Spanish Conquistadores went to South America; the Spaniards expressed their surprise at the silver being carried over the high mountain passes by what they described as "Peruvian sheep". The llama is the only beast able to carry burdens in these cold high altitudes and it is of great value even though its load is limited to under a hundredweight and its journey to about fourteen miles a day; in addition the llama provides meat, milk and wool. The alpaca's sole use is for wool production, and its fleece is generally both longer and finer than the true llama's. Both tame forms are known in various colours, principally white, brown or black, or parti-coloured.

Finally we come to the elephant, by far the largest land animal in the world today. It is the outstanding beast of burden of the Indian and Burmese jungles and it is of great economic importance, doing all sorts of jobs that are never likely to be mechanized. Strictly speaking, however, the elephant is not a domesticated animal, for although elephants are from time to time born in captivity, man is normally dependent on wild animals, caught in annual drives and trained with the skilled and willing assistance of experienced older beasts. Even so, this chapter is obviously where the elephant should find a place. It is probably at its finest on timber work, hauling, pushing and generally handling the teak and other valuable hardwoods; but for sporting work also, carrying the *shikari* and his loads, and beating the jungle for tiger or leopard, it has no equal, and it still takes a large part in the rich ceremony of the East.

The use of elephants goes back for many centuries in India, and they enter largely into Indian and Burmese art. Hannibal's employment of elephants for military purposes in the 3rd century B.C. is well known, but there is still disagreement as to which elephants were used; at first it was assumed that for obvious geographical reasons they were of African origin, but when it seemed that the African elephant could not be trained to work, they were assumed to have been Asiatic. Further research indicates that Hannibal's elephants were quite probably African after all. There is little doubt that the first tame elephants seen in the eastern Mediterranean area had come from India, though the Egyptians and Romans may have turned to Africa for subsequent supplies. But the numbers which they trained are quite staggering: in the famous procession in honour of Dionysus, described in detail in another chapter, the lead was taken by no fewer than ninety-six elephants drawing chariots in teams of four. There are not many more than that number in all the Zoos of western Europe today!

Elephants need a great bulk of food for efficient working; in India it is often reckoned that an animal can work for some eight hours and rest for eight hours. The African species certainly yields less easily to man's domination, but the real snag comes a little later; it seems virtually impossible for it to collect and eat enough foliage and still leave time for a reasonable working day. In the Belgian Congo a fine herd of elephants has been trained, but economically the experiment was not a success, though of great biological interest, and it has now been dropped. It was also very wasteful of animal life, for a number were lost for every one trained.

Elephants can be taught to do a great variety of things: useful, as in the jungles; ornamental, as in the traditional ceremonial processions in India; and interesting or amusing, as in their performance in Zoos or circuses. It is true that elephants yield ivory, one of the most precious of all animal products, but ivory comes from wild tuskers—after they have been killed. The elephant thus shares with the donkey the distinction of being brought into the service of man purely and simply for what it can do.

3. COMPANIONS AND ASSISTANTS

THE last main section of fully domesticated animals can be described, very roughly and rather inadequately, as companions of man. Of these the dog is the doyen and it is quite truly the companion of man in a thousand and one activities: the association began in Neolithic times and has continued unbroken ever since. Dog skeletons have been found in Neolithic kitchen middens in Denmark and their remains are the only sign of domestic animals in the settlements of the Stone Age hunter-fishers near Scarborough, dated 7000–8000 B.C.

Professor Wood Jones, the well-known anatomist, suggests that the dog was domesticated in several stages: first a camp-follower, then a hanger-on; gradually it became more dependent and finally it was man's companion. There has been much argument about the origin of the domestic dog which we know today in a fantastic range of colours, sizes and shapes; many authorities have held that these dogs are the result of domesticating several different wolves, jackals and wild dogs, but modern research supports the view that they are all in fact descended from the common wolf. The modern dog interbreeds fairly readily with the wolf and Aristotle records such a hybrid in the 4th century B.C.

In quite early times the dog was venerated, especially in Egypt and Ethiopia. It was considered a desirable goal for the human spirit after death, and therefore largely on religious grounds the Israelites were taught to regard it with abhorrence, though there was also a hygienic reason, the dog of those days being largely dependent on scavenging for its food.

An Indian ivory carver at work

Exclusive News Agency

Superb craftsmanship in leather

John Mason

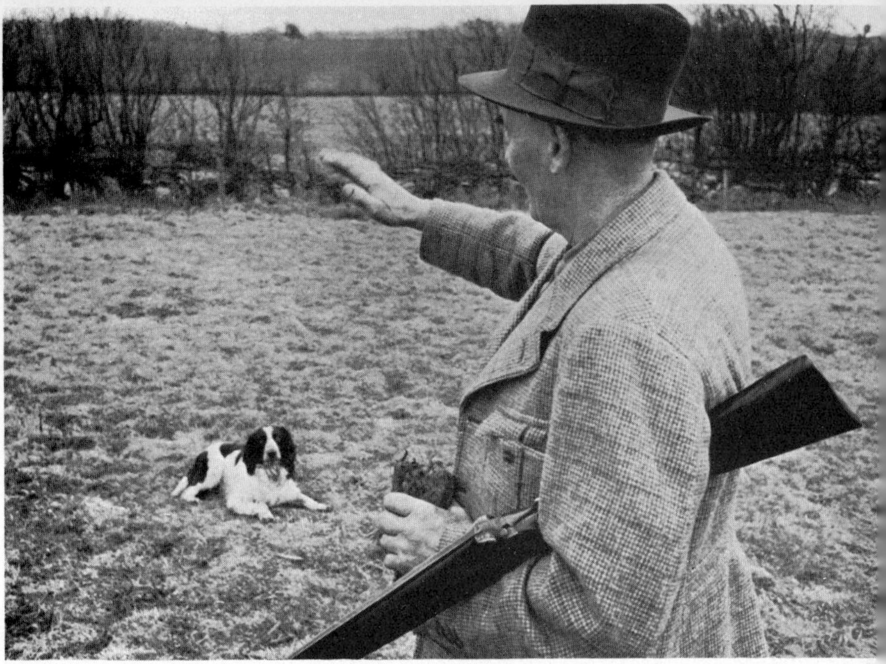

Picture Post Libra

A spaniel being trained as a gun dog

The huntsman calling in the hounds after they have lost the fox

Picture Post Libra

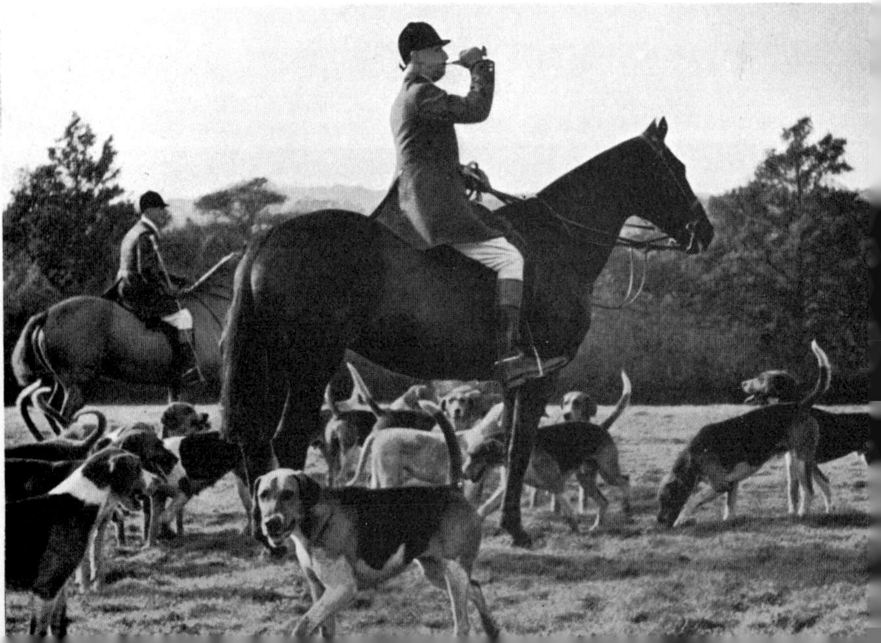

For several thousand years the dogs were very like their wild forebears and were quite unspecialized, but by the third millennium B.C. the Babylonians had developed a mastiff type which is shown in art of that period. Egyptian monuments of at least one thousand years earlier show dogs very like the modern greyhound and others like the Aberdeen terrier. The world-wide distribution of the dog must have taken place very early and special forms were developed in almost every country.

Mastiffs had probably been brought to Britain by the Phoenicians, for Julius Caesar found them here: dog-breeding seems to have started early and by the time of Augustus hunting-dogs were being exported. Alfred the Great took an interest in dogs and he made plans for the training of dog-keepers and falconers towards the end of the 9th century: a century later we read of sheepdogs at work. By the 13th century dogs were being used for poaching, but not long after that the villagers complained that rabbits were damaging their crops because dogs had been forbidden and they were again allowed to keep dogs—but they had to be small ones and able to pass through a certain size of hoop! In the Middle Ages the bloodhound was already being used for tracking, especially in the Scottish borders, and it was ruled that anybody obstructing a hound on the trail should be held as an accessary.

The dog has proved an almost incredibly plastic animal and it has manifested a greater range of sizes and shapes, and been developed for a wider variety of purposes, than any other beast that serves man. Dogs vary in size from toy breeds of 4 or 5 lb. to Great Danes of over 9 stone. Some two hundred more or less distinct breeds are recognized, and most of these are man-made, in so far as they have been deliberately selected for certain characters. Many of them are nothing but freaks, though their fond owners would not agree with this description, but such "sports" as the Pekinese and the bulldog could never survive without human assistance.

It is quite impossible to list more than a small fraction of the myriad jobs which man finds for his dogs. He uses them for defence of stock and property, for life-saving and for guiding the blind; for sport—greyhounds, foxhounds, dachshunds, terriers, setters, spaniels and retrievers all working in different ways; as working dogs, mostly for herding sheep and drawing small carts; for meat and fur—this was largely the chow's original purpose. Lastly, there is a big group of ornamental dogs, and a very large number of nondescripts, which can be said to give only one thing—companionship. A very old book on dogs suggests quite a novel reason for keeping them in its reference to "the smaller ladyes poppes that bear away the fleas and dyvers small fowles". Nowadays we probably find it cheaper to use D.D.T.!

The domestic cat has a very different history from the dog's. Some authorities consider that it first appears near the beginning of Egyptian civilization, but the first positive evidence is for the 16th century B.C. However, it is probably the only truly Egyptian contribution to domestic stock and it also appears to be the only one which does not come from a more or less gregarious species; the ancient records suggest that some other species of cat may well have been tamed, but they have been lost long since. The chief ancestor is undoubtedly the north-east African wild cat, but other wild cats may well have added to the general stock, including the European wild cat, which is definitely known to hybridize with domestic cats.

It seems likely that the need for mouse and rat control was a big factor in bringing the cat into domestication in Egypt, but we shall never know for certain. As Professor Zeuner points out, it may well be that for some time the cat was hanging around human habitations catching small rodents and being tolerated because of its usefulness without actually being bred in captivity. Egypt was for millennia the granary of the ancient world and some old records show the cats engaged in battles with the rats; it has also been suggested that the Egyptian priests noted the association of rats with bubonic plague and that the deification of the cat might have been an astute move on their part to get the rats controlled. The cat certainly was a sacred animal in Egypt and for this reason, presumably, was anathema to the Hebrews, who omit it entirely from their records. It has few mentions in other ancient records, which is perhaps not surprising in view of the fact that it did not leave the Nile valley until about the 1st century B.C.

The Romans took the cat with them as they colonized parts of Europe; the remains have recently been found in a house in Kent of a cat which died during the 2nd century A.D.

and these show it to have been of the domestic type. It seems, however, that it was confined to the rich and did not become generally distributed until about the 9th century. Later on cats came to be associated with witches and suffered, if anything, worse treatment even than they; it has been suggested that some of the outbreaks of plague in western Europe, especially in England, are correlated with these drives against the cat, after which the rat's numbers got out of hand.

The anatomy of the cat has changed very little, but a fairly wide range of coat and colour has been developed: these are all the result of sports which have been deliberately selected and established. The geographical names attached to these varieties mean little: the Manx cat is often considered as introduced to the Isle of Man and there is no evidence that the Siamese cat first came from Siam. This semi-albino cat is the subject of much argument as to its origin: it has been known in the West for less than a century, but Dr. L. Harrison Matthews has now drawn my attention to an illustration, much older than this, of a cat extraordinarily like a Siamese. This is Plate I in the English edition of *Travels through the Southern Provinces of the Russian Empire in 1793-4* by P. S. Pallas, published in London in 1802. But the greatest efforts are needed to keep any breed pure, for, given half a chance, the cat is out on the tiles and everything gets mixed up again. The cat's food habits have changed in captivity and it is a little hard to explain its fondness for fish and milk, also its intense dislike of getting wet. The domestic cat reverts to the wild state and becomes quite independent more easily than any other species, as the Australians and New Zealanders know to their cost.

Today cats of almost all shades are found all over the world and, for the most part, they are companionable pets. Once in a while a cat makes news by acting as a retriever, but this is nothing new, for in Egypt cats were used regularly in this way. Some warehouse cats have to earn their keep; many household cats do a little mouse-hunting on the side; some are unbelievably pampered pets, but the great majority are just pets, though, it must be said, they are generally pets with a greater allegiance to a place than to a person, in contrast to most dogs.

The ferret is the only other carnivore that has been domesticated—apart from the species now under development for fur production—and it is an albino form of one of the polecats. Opinions about its origin differ: various authorities suggest respectively the European polecat, an Asiatic species and one formerly found in parts of North Africa. The ferret is known to hybridize with the European form and produce a variety known as the polecat-ferret, but the most probable ancestor seems to have lived in Asia. It is possible that the ferret was known to the Greeks and kept in their houses to destroy vermin, but it is rather hard to sort out the names they apply: the word mostly used is one generally applied to the weasel, but the context certainly suggests the polecat, perhaps before it was actually domesticated. The Romans used the ferret and Pliny, writing nearly 2,000 years ago, says that they were used for rabbiting. Strabo reported that when the Spanish people were suffering from a plague of rabbits they applied to Rome for help and were sent a supply of ferrets. They were instructed to put the ferrets down the holes and bolt the rabbits into nets, precisely as is done today in many parts of this country.

The ferret shows little change after this long period of domestication but remains a true albino with pink eyes; the polecat-ferrets generally favour the polecat in colour and size. Ferrets are still used for rabbit catching and have been taken to many countries to do this. In Great Britain escaped ferrets do not survive very long, but in some warmer countries, such as the Canary Islands, they have become feral and developed into distinct local forms; this suggests that the ferret may have first come from a warmer climate. The ferret has recently found a new use as a laboratory animal, for it is one of very few mammals capable of being infected with the human influenza virus.

Although countless varieties of birds are kept as pets in almost every part of the world, only five of these can be classed as domesticated and two of these are of minor importance. The canary is, of course, a firm favourite: it is derived from a finch found only in the Canary Islands, off the north-west coast of Africa, and it became distributed in Europe about the middle of the 16th century, though for some time the Spanish had tried to maintain a monopoly by exporting only males. It is recorded that a Spanish ship carrying many canaries was wrecked on the island of Elba and that the birds colonized the island, forming a stock from which birds were sent all over Europe. Most of the domesticated varieties of canary are much more yellow than the wild bird and they have been developed along very specialized lines known by such names as Border, Yorkshire and Lizard. Only those in "the fancy" can appreciate the really fine points, but to many thousands of people the canary is a delightful singing pet.

The budgerigar became domesticated much more recently; it was, in fact, brought to Great Britain from Australia by Gould in 1840, and it is the only animal from the Antipodes to become domesticated. It is one of the smallest of the parrot tribe and a bird that is comparatively easy to keep in temperate climates. Soon after its introduction to Europe colour varieties began to appear and within thirty or forty years yellow and blue forms were fully established. The past few decades have produced a wide series of colour forms through the careful selective breeding from mutants, and some of these are very beautiful and command fancy prices. One other small ornamental bird deserves a mention here, though it is probably known only to bird fanciers: this is the bengalee, a domesticated mannikin (a small finch-like bird), whose chief value is that it is a perfect foster-mother for hatching the eggs of small cage birds that will not do this themselves.

The domestic pigeon is descended from the rock dove found around our coasts as well as in many other parts of Europe and Africa, and it has a long history. It figures on monuments as far back as the earliest dynasties in Egypt and the first record of its being used as a table bird is in the Fourth Dynasty, some 4000 B.C. There is little doubt that it was first kept for food and, arising out of this, used for sacrificial purposes; then, later, for carrying siege and business messages; finally for showing and racing. The ancient Egyptians may possibly have used pigeons for message-carrying and the Greeks knew this could be done, for one Taurosthenes used to employ a pigeon to send home news of the Olympic Games. Of the few historical records of the pigeon post the best known concerns the siege of Modena in 78 B.C., when messages were flown in from the commander of the relieving army. The world-famous firm of Reuters started off with a pigeon post, and even today, in spite of the wireless, carrier pigeons still have their uses.

The Romans kept pigeons on a tremendous scale and their special houses often held 5,000 birds. In mediaeval Great Britain it was also the custom to erect vast dove-cotes: one still stands at Sibthorpe, in Nottingham, built over 600 years ago; it is thirty feet in diameter and sixty feet high, and at one time had 1,280 nesting boxes. In all these cases of mass production the birds were semi-wild and used only for food, manure being a useful by-product. Today there are over two hundred distinct breeds, many of them highly artificial and quite unable to survive alone, though the carrier pigeon remains fairly like its wild forebear. The London pigeon is the result of these birds escaping from full captivity and becoming semi-feral: the more striking abnormalities of colour and shape are soon lost.

Pigeons have many uses in addition to those already mentioned. They are trained to perform in circuses and to smuggle drugs ashore; until it was made illegal they were used in trap-shooting from boxes. They are useful laboratory animals and are the standard bird for dissection by zoology students. But today they are kept for three main purposes—for showing, for racing and as pets.

The Barbary turtle dove was also domesticated far back in antiquity; it is mentioned frequently in the Biblical record, where it is often associated with gentleness and affection, as it indeed is throughout the ages. It was also regularly used as the poor person's sacrifice and in certain particular ceremonial offerings. However, it has never been of great importance and is seldom seen today except in aviaries.

In addition to these five fairly small birds associated definitely with houses, even if not actually kept as pets, there is a group of birds which have been domesticated, or at least brought under some sort of control, for ornamental reasons, to grace the estates of their owners. The mute swan is perhaps the best example of this class. It was formerly indigenous in parts of East Anglia, but was brought into semi-domestication in the 11th or 12th century and ceased to exist as a wild British bird. Up to the 18th century all swans were owned by the

Crown or under licence granted by the Crown, with the observance of a strict code of statutes and customs. This control has largely been dropped, but the King's Swanmaster still organizes the annual swan-upping on the Thames. In some parts of the country the swans have more or less reverted to the wild state and this even applies to the famous Abbotsbury Swannery, where from 200 to 500 pairs breed every year.

The peacock is really the swan's opposite number on land, but it has been domesticated for very much longer. It came to western Asia from India very early and Solomon knew it about 1000 B.C. It seems to have reached Athens about 450 B.C., probably from the isle of Samos, where peafowl had been kept as temple birds. Alexander the Great may have brought some back from the East with him, but these were by no means the first arrivals in Greece, where they were always rare and expensive birds, though the Romans bred them in numbers, largely for the table; peacock was always a fashionable and festive dish. Peafowl were kept semi-wild on small islands off the Italian coast, but on the mainland they were kept rather like fowls. Later on the peacock was found quite useful as a sentinel at night, but this very noisiness makes it a most unwelcome bird in built-up areas. The normal tame bird is exactly like its wild forebear, but a few colour varieties are known, including the black-shouldered and white peafowl.

Some of the pheasants and partridges might perhaps be classed with the peafowl as ornamental estate birds, especially the golden, silver, Reeves's and Lady Amherst pheasants and the chukor partridges, but they are certainly not domesticated birds. The common pheasant brightens up the woodlands and fields, but it was introduced and established purely as a sporting bird.

After the ornamental birds it is perhaps convenient to deal with a small group of fish, though for them "domesticated" may hardly seem the most appropriate word. In Europe the Romans had started the fashion for keeping fish in tanks and ponds, especially sea fish; the chief object was to provide food for the table, but some enthusiasts would never allow their particular pets to be eaten. Later, the Romans may have taken an interest in freshwater fish, though this is not very clear, but we can perhaps regard the carp ponds of mediaeval Europe as the successors of the Mediterranean pools. The carp, incidentally, is not a native of western Europe but comes from eastern Asia. Several of the forms now found in our ponds and lakes are mutants which have been deliberately propagated.

Away in the East the Chinese had been developing the goldfish. Tradition has it that this cult is of great antiquity, but the first quite definite reference so far found is dated about A.D. 450. The goldfish is derived from the golden carp, a native of China, and many ornamental forms are kept today, in both ponds and aquaria; some of these, with veil tails, telescopic eyes and other abnormalities, are nothing but monstrosities that could never survive in the wild. The goldfish reached Great Britain about the end of the 17th century—1691 seems to be an authentic date—having come in easy stages via Batavia, Mauritius and St. Helena; in those days of sailing ships it was not easy to transport fish long distances. The paradise fish originated in the East and for a long time was considered a separate species, but it is now known to be a long-domesticated variety of *Polycanthus*, a fish native to China and Cochin China.

The past two decades have seen a big development in tropical fish-keeping; this can be compared fairly closely to the keeping of assorted cage-birds, but special varieties and hybrids are being developed which now almost qualify for the description "domesticated".

Several forms of trout and salmon receive a great deal of attention and their numbers, especially in the case of trout, are greatly increased by artificial rearing, but they cannot be considered in the least domesticated. The position of the oyster is perhaps more open to argument; it has always been gathered to some extent from natural beds, but before 90 B.C. the Romans were laying down special beds where the oysters could be protected and fattened. Modern methods of oyster culture date from French work in the mid-10th century; France is still a very big producer and at Arcachon the harvest comes to 650,000,000 in a good year. Two more marine shellfish come in somewhat the same class as the oyster: in the Far East a kind of oyster is used for the production of "cultured" pearls and in parts of Europe the mussel is grown in conditions in which a quite distinct form has developed.

In recent years an entirely new class of animal has arisen, to meet the needs of scientific research. Most of the animals coming into this class had already been domesticated, but several of them are now used primarily for this new work. The laboratory animal is, in fact, typified by the guinea-pig, which has given its name to the whole class of animals under test, human or otherwise. The first use of the word "guinea-pig" in this connection seems to have been in the late 1930s, but two previous human applications of "guinea-pig" were current a century ago, meaning a person whose standard fee was a guinea and a nominal company director.

The name suggests a West African origin but, as in various other cases, "Guinea" mainly implied an overseas origin; the guinea-pig, or cavy, actually came from South America, where it had been domesticated by the Peruvians before the Spanish conquest. The wild form is a small greyish rodent that lives in burrows in the dry sandy area along the Pacific coast. Though often reported to have been brought to Europe in the middle of the 16th century there is little or no evidence that it arrived until at least a century later, and it came to England about 1750. In the first place the Incas seem to have used them for food and for sacrificial purposes; even today they are eaten in various parts of the world, including West Africa, where they are known as "Lagos mice". The German name *"Meerschweinchen"* means "little sea pig" or "little porpoise".

Several colour forms were already established when the Spanish reached Peru: a range of colours, both plain and mixed, are known today, and there are also various types of hair, including long, short and whorled. Until well into the present century they were little other than pets in Great Britain, but then they were found to be very suitable animals for the many laboratory tests demanded by modern medicine and pharmacy. In the wild form only one or two are born at a time, but the domesticated cavy has litters of up to eight or even ten; this was one of the factors that made them such suitable laboratory animals, for they could be more or less mass-produced to a close pattern, especially in size.

The laboratory fashion has now changed considerably, and rats and mice are used to an increasing degree; their breeding has been thoroughly standardized and they can be produced much more cheaply than a guinea-pig. In addition, their short gestation period, of only about three weeks against ten weeks for the guinea-pig, is an advantage.

These smaller rodents had previously been domesticated as pets, the white and parti-coloured rats from the brown rat (*Rattus norvegicus*), and the white and coloured mice from the house mouse (*Mus musculinus*). Both of these pests had come to Europe from the East, the brown rat early in the 18th century and the house mouse perhaps a little earlier.

The golden hamster is a very recent recruit to the army of pets and laboratory animals and it is a beast with a fascinating history. It was first recorded in 1839 from Syria, once again early this century and then next in 1930, when a family was taken alive: the whole of the prodigious number of golden hamsters in captivity today have descended from part of this family of twelve. It has become so popular as a pet that Golden Hamster Clubs have been formed; it is also a well-established laboratory animal, thanks partly to its large litters and its short gestation period of only fifteen days.

A number of other animals, including the rabbit, are used for laboratory work, but only one other vertebrate deserves special mention—*Xenopus*, the clawed frog. This strange South African frog, more fully aquatic than any other, proved to be of great value in pregnancy determination, but in the last few years other frogs and toads have been found almost as useful in this way.

Among the invertebrates, the little fruit-fly *Drosophila* has been the subject of very extensive research, particularly in the realm of genetics, and it can certainly claim to be fully established as a laboratory animal, with a number of stable forms quite unlike the original wild insect. The wingless form, for instance, has been especially developed for feeding to minute tropical amphibians and humming-birds in zoos.

Finally there are a few animals which man has enlisted for various jobs, but because they have not so far been established as breeding forms in captivity they can hardly be classified as completely domesticated. The elephant, already mentioned, might be placed in this category of "allies". Strangely enough, the Primates are one of the few major groups which have not produced a fully domesticated form; organ-grinders have—or always used to have—their monkey assistants, and zoos and circuses may have their chimpanzees' tea parties, but the nearest approach to domestication is probably found in the Malayan region where the pig-tailed monkey is regularly enlisted and trained for coconut-picking. These monkeys, like elephants, are not bred in captivity, but caught about half-grown and trained to climb the palms and twist off and throw down the nuts. A very interesting development is their use for botanical collecting. Anyone who has tried to make a serious collection in the wet tropics knows just how difficult this can be, for the flowers and leaves of many of the trees and tall climbers are far beyond human reach, but foresters in Malay have been able to train a coconut-picking pig-tailed monkey —after about a six months' course—to be an expert plant-collector, knowing at least a dozen commands which are sufficient to direct most of its activities. An officer who trained several reported of one that it had collected from twenty-four trees of over a hundred feet in height in one day, a task that could only have been accomplished otherwise by cutting down the trees.

The cheetah, often reckoned to be the fastest mammal in the sprint, has for many centuries been used for the chase, especially by the Indian Princes. Best described, perhaps, as a cat with several doggy characteristics, it can become completely tame and be trained after capture as a young adult; in spite of this, however, it has never been known to breed in captivity anywhere, even in its native land.

The civet cat has a claim to mention here, for it is kept captive in several parts of Africa—principally Nigeria and Abyssinia—solely for the secretions of its anal glands, from which the musk-like scent of civet is obtained. The secretion is removed with a smooth bone spoon and if the civets are fed properly they should yield from half to one ounce of crude scent per week. The civets are wild caught and very seldom breed in captivity. Not everybody appreciates this scent, for in the Gold Coast the name for the civet in a number of languages is, literally, "stinker".

For centuries—perhaps for millennia—the Chinese have trained and used the cormorant for fishing, first tying a cord around the neck to stop it swallowing its prey, though the best birds can be trusted to fish without any such control. Fishing with the aid of cormorants was introduced to England in the time of the Stuarts and it also had some following in other parts of western Europe, but the cormorants were birds reared from young taken out of the nest; in China, on the other hand, the tame cormorants lay eggs which are hatched under domestic hens, and they have some claim to be considered as domesticated. It is interesting to note that the Chinese have been able to enlist as an ally a bird which, in Europe generally, is regarded as a pest or, at best, a competitor for the fishing.

The art and sport of falconry is almost as ancient and far more widespread than the use of cormorants. In the British Isles it reached its peak in the Middle Ages, and Shakespeare refers to it many times. When Hamlet said "I am but mad north-north-west: when the wind is southerly I know a hawk from a handsaw" he was comparing the peregrine falcon, prince of hawks, with the heron or hernshaw, in those days its classic prey. In Persia and India, however, it has never lost its importance. In western Europe recent years have seen its revival, with the use of a variety of Raptores—golden eagle, goshawk and a number of smaller kinds, as well as the peregrine, which had the honour of being conscripted during World War II for keeping airfields free of pigeons. Around the coasts, however, the wild peregrine had to be put back on the vermin list to protect carrier pigeons engaged on all sorts of war jobs.

It is now quite impossible to envisage human life without domesticated animals. The Laplanders depend on their reindeer and the humblest Africans on their hunting dogs and cattle, just as the sophisticated flat-dweller takes in her daily bottle of T.T. milk, but they seldom if ever think of their very distant ancestors who began to make their various modes of life possible by enlisting the help of animals. The groundwork in domestication was more or less completed several thousand years ago and few animals of first importance have been added to the list in our own era. The work, however, is by no means finished and the expert will long be kept busy breeding just the right pig, cow or sheep for man's latest needs.

Part Four

Other Associations Between Animals and Man

1. ANIMALS IN SPORT

ALL out-of-doors pastimes are known generally as sports and a large proportion of them are connected with animals in one way or another; to some people, in fact, sport consists of nothing other than going out to catch or shoot some animal, and it is this sort of person who says, "It's a fine day: let's go out and kill something." This connection of animals with sport is now more or less universal; it is certainly of great antiquity and there can be little doubt that it all started with hunting. In the early days man hunted to live; as agriculture developed he went on hunting to get all or part of his meat supply as well as to get variety of food, but over large areas of the world today man is still entirely dependent on hunting for his meat. From killing for food to killing for sport as well as food is not a very big step, and the genuine poacher is but showing the primitive urge to hunt, coupled generally with a feeling that stolen waters are sweet, though it must be admitted that most modern poachers are merely mercenary.

Some sports are connected very intimately with the domestication of the dog and the development of specialized breeds: sporting dogs are portrayed on ancient Assyrian and Egyptian monuments and these peoples already had the recognizable forerunners of mastiffs and greyhounds, which they used in hunting. Falconry and hunting with cheetahs were practised by the nobility in the old Indian civilizations and animal spectacles were provided on a gigantic scale by the rulers of ancient Rome. It is a far cry from those days to the present.

Some sports have developed and had a period of popularity before dropping out or being banned as public tastes have changed, but with many sports the only difference is in the methods used: as, for instance, in hunting, the bow was successively replaced by the cross-bow, flintlock, cap gun and, finally, the precision rifle complete with telescopic sight.

The sports and pastimes, both old and new, involving animals are so many and so varied that it is perhaps convenient to consider them in four main groups, according to the relation between man and the animals concerned in each of them.

(1) Where the object is to catch, and generally to kill, the animal as quickly and neatly as possible—the hunting, shooting and fishing.

(2) Where animals fight each other: as in cock-fighting and bull-baiting.

(3) Where the animal is used for man's pleasure, without getting hurt: as in horse-riding, horse-racing and greyhound-racing.

(4) Where man's attitude is largely altruistic: as in bird-watching and aquarium keeping.

Some possible effects of hunting were considered in earlier chapters. The red deer, the most important British animal of the chase until a few hundred years ago, has now been so reduced that it provides sport in only a very limited area of these islands. In Scotland it is stalked and, ideally, a high sporting code is observed, though unfortunately there is as yet no law to protect the deer adequately in seasons when it is unsuitable for killing as meat. In a few parts of Great Britain the stag is still hunted—principally on Exmoor—and there are also one or two packs of buckhounds, for hunting the fallow deer, the best known of these being the New Forest Buckhounds. These packs hunt to kill, but there are other packs which hunt a stag specially caught up and released in a convenient place; when the stag gets tired it just stops and faces the hounds, having learnt that it will be caught up and used again. After a stag has been hunted a few times it gets to know a route and follows it regularly; a stag has also been known to run a small circle and come straight back to the horse box in which it had been brought! Hunting the

Victoria and Albert Museum

The Mogul Emperor Akbar hunting blackbuck with cheetah

Students at the Field Study Centre of Flatford Mill

carted stag is, therefore, a blood sport made bloodless; the final stage in this direction is to substitute a horseman laying a scent trail, thus making what is known as a drag-hunt.

Fox-hunting is the subject of many bitter attacks, but it cannot be denied that in many parts of the country foxes are too numerous and must be kept in check, and, further, that shooting them—the only practical alternative—can be both cruel and not always effective. Fox-hunting varies infinitely from the highly fashionable gatherings in the Shires, with large and very well-mounted followings, to the strictly utilitarian hunts in the Fells, where the packs are mixed and the fields which follow on foot are a cross between steeple-chasers and rock-climbers. The hill fox has a good chance of escaping and by no means every run leads to a kill.

Otter-hunting is very different from fox-hunting and, to many people, much less justifiable; the otter may be a nuisance locally, but the damage it does is often overrated and it is seldom common enough to justify hunting for purely control purposes. This is a much less expensive sport than fox-hunting and done entirely on foot. There are various types of hounds but they are mostly rough-haired and, of course, they hunt by scent alone. There is a high code of sportsmanship in otter-hunting and in most places a blank day is more common than a kill.

The hare is another British mammal which, like the fox, is in no danger of extermination, and it is probably hunted in more varied ways than any other animal. It is the classic prey for the old-time poacher working with his net and his "long dog" or lurcher. It is more legitimately taken at coursing meetings where greyhounds are released, two at a time, after more exciting prey than the mechanical toy that most of them spend their lives chasing. Coursing hares that have been trapped alive and then released has nothing to commend it. Hares are also hunted by several kinds of true hounds, which work by scent and not by sight, as do the greyhounds. Harriers are akin to foxhounds and are followed by a mounted field, whereas beagles are a shorter, slower hound followed on foot. But more hares are shot than killed by all other means together.

The ubiquitous rabbit is a pest to be destroyed at all times and by all means; great numbers are taken by professional trappers, though they will never exterminate it, but the rabbit is also the backbone of many a small man's shoot and provides good sport, whether shot as it leaves the last patch of standing corn in front of the self-binder or as it is bolted from its hole by a ferret. The rabbit, too, is the standard quarry for the goshawk, though it is unlikely that more than two or three are at present being used for falconry in this country.

In the really big shoot, however, rabbits hardly count, for this is based on the gamebirds, and Great Britain has—or once had—many estates world-famous for their tremendous bags. In the early days of this century the daily bags of pheasants, partridges and grouse quite frequently reached four figures, but those days are almost certainly gone for good, especially those based on the artificial rearing of pheasants; that is perhaps as well, for they were so completely organized that almost the only thing demanded of the sportsman was that he should shoot accurately with the gun refilled and handed him by his loader. The final development in artificiality is the syndicate shoot in which the members may be town-dwellers scarcely able to identify the birds they shoot and with no knowledge of their habits. They can hardly be compared with the solitary sportsman who walked up partridges or shot them over pointers with his muzzle-loader; the field-craft demanded of him should surely be an integral and necessary part of all field sports.

There will always be argument about the most difficult shot: many say it is the very high pheasant with the wind behind it, but the man who successfully outwits wood-pigeons and comes home with a full bag has demonstrated a mixture of skills of which aiming the gun is only a part; while the wild-fowler needs, in addition, a precise knowledge of tides, winds, topography and, not least of all, the habits of his very elusive quarry. His, perhaps, is the greatest satisfaction of all. The side effects of intensive game preservation in Great Britain have been discussed in an earlier chapter.

Few countries are without birds or animals generally considered as "game"; such animals are a little hard to define, but the term implies that their killing demands skill and therefore provides sport, and also that they are good to eat.

One large group of birds is classified as "gamebirds"—the pheasants, guinea-fowl, grouse and so on; most of them are strong runners and swift fliers, once they take to the air. Of birds hunted for sport the majority, though by no means all, are gamebirds. Duck-shooting in Canada, snipe-shooting in the Burma paddy fields, sand-grouse-shooting on the Indian plains and partridge-stalking in the Sinai hills—these are all world-famous forms of bird-shooting.

Big-game hunting was a form of shooting that reached its peak early this century. The African plains were its headquarters, but most of the less densely populated parts of the world had their specialities, from the highlands of Central Asia to the jungles of Burma and Malaya and the mountain forests of the Caucasus. Except for the resident this was always a fairly expensive sport, but today its cost is prohibitive for all except the tiny minority; for these there exist in East Africa organizations which will undertake everything connected with the trip so that all the big-game hunter need do is to shoot accurately—and pay! The areas available for hunting are continually being reduced and the licences made more and more restricted and expensive. Fortunately there seems to be a decreasing desire to kill and there is certainly little incentive to bring back large collections of trophies, except to museums.

Pig-sticking—the hunting of wild pig by horsemen armed with spears—is practised for the most part in India and it has all the elements of a first-rate sport. It takes place in wild, open country, it involves the use of fast, intelligent horses and the crafty wild boar is by no means a defenceless quarry. In addition it has the competitive spirit, for pig-sticking is generally organized in heats of three or four riders and the man who wins "First spear" scores the points. The object is to spear the pig through the heart and it is thus killed quickly and mercifully. The use of cheetahs for hunting gazelles and blackbuck and the art of falconry are two other sports which originated in the East, though the latter is now more widespread. These have already been mentioned in connection with domestication.

Fishing, the last of the pastimes involving the catching or killing of animals, is by far the most popular of all. In western Europe and North America there must be several dozen fishing rods for every shotgun or rifle, and angling is as much enjoyed by the salmon fisherman casting his fly on the most expensive beat of the Tay as by the small boy catching a few roach or carp in the village pond. Like most other sports, fishing becomes increasingly artificial and in many famous fisheries the trout are reared and then introduced to the streams or lakes; but however much is done for him, the angler still has to find, hook and land his fish and that, after all, is the essence of fishing. Big game is not confined to the land; well-organized trips for "big-game fishing" are now available in many parts of the world—to those that can afford them.

We turn now to the second category of sports, those in which the animals involved get hurt or killed by one another. With very few exceptions these sports have been given up by civilized peoples and not one of them is practised legally in Great Britain today. The ancient Greeks and Romans, with their tremendous spectacles, started this type of sport, but their activities will be considered more particularly in connection with the development of menageries.

Bull-fighting seems to have been started in Thessaly by the Greeks, but at first it was more like a modern rodeo in that the bulls were chased by horsemen who then leapt on their backs and brought them down by their horns, breaking their necks. Modern bull-fighting is full of pageantry and tradition, and it involves a high degree of skill, but it is distasteful to the majority of Europeans, even though it is still a national sport in south-western Europe and parts of America. It inevitably causes the bull pain and to most sportsmen it is also objectionable because the bull never has a chance to escape; whether it fights bravely or refuses to fight at all it is quite certain to be dragged out dead.

Bull-baiting and bear-baiting are pastimes which had their heyday three or four centuries ago. Bear-baiting was always an expensive sport and one indulged in by the Court, so it is not surprising to find that it was an ancient custom to supply one or two bears as an annual due. A bear garden was, literally, a place where bears were baited by powerful dogs. In the 16th century people were already complaining about the cruelty involved, and bear-baiting more or less died out in the middle of the following century, though it is recorded that a bear-baiting was advertised, perhaps for the last time, in 1716. The chief reason for bear-baiting seems to have been nothing other than the baser instinct of man such as called for the savagery of Roman games; but bull-baiting was slightly different in its origin, for it was firmly believed that bull's blood and flesh were dangerous as food in their natural state. They had, therefore, to be prepared before the bull was killed, and the bull was harried in the belief that this made the blood thin and the flesh edible! There were, in fact, local laws decreeing that "no butcher kill a bull till baited", and there are records of butchers being fined on this account early in the 17th century. But this was only an excuse, and the real reason for baiting bulls was that the common people liked to see it done.

Bulldogs were developed largely for bull-baiting, but the modern breed would probably be little good for it; dogs very like the modern Staffordshire bull terrier are also shown in 18th-century pictures of the sport. Bull-baiting and dog-fighting were both made illegal in 1835, but a small group of enthusiasts tried to carry on for a few more years before being put into prison for breaking the law.

Of badger-baiting the less said the better. It was a barbarous pastime which today would not be tolerated anywhere. The poor badger was generally fastened to a stake, often in a half-starved condition, and set upon by dogs, with much damage to both parties. This should not be confused with badger-digging, a procedure sometimes employed nowadays quite legitimately for catching a live badger or killing one that has become a pest.

Cock-fighting is an even older sport than bull-fighting and it was popular in the early civilizations of India, Persia and China. It seems to have been introduced into Greece in the time of Themistocles and it became a favourite sport in Athens. Pinder wrote about cock-fighting in some detail as far back as 470 B.C.; in those days cocks were bred specially for fighting and it was usual to feed them on leeks and onions. This sport was taken up enthusiastically in Rome towards the end of the 2nd century B.C. and very heavy betting took place. Owners even went so far as to dope the birds with stimulants. Special breeds of domestic fowl were generally used, but the

Romans also trained other gamebirds such as rock partridges and quails for fighting.

From Rome, cock-fighting—or cocking, as it came to be called later—spread to most of western Europe and it is possible that it was brought to England even before Julius Caesar arrived. It was very popular in the Middle Ages and in the time of Henry II was particularly the sport of schoolboys, who even expected special allowances to buy and train their birds. Cock-fighting was perhaps at its height at the beginning of the 19th century. Special breeds had been developed, most of them carrying wickedly long and sharp spurs; steel spurs were even fitted for some fights. Unless the birds refused to carry on a fight, which was unusual, most combats were therefore to the death. In Wales it was apparently the custom to stage fights in churchyards and even in church itself, but in the cities special cock-pits were built; the name 'cock-pit' survives chiefly as the pilot's cabin in an aircraft, but in a few places it marks the former site of a literal cock-pit.

Cock-fighting became a very fashionable and expensive sport and it is recorded that the stake for a match fought in 1830 was 5,000 guineas, with further heavy bets on the individual fights. Soon after this public opinion began to turn against this type of sport and cocking was made illegal by an Act of Parliament in 1849, but it is still allowed in some parts of southern Europe and it remains a favourite sport in parts of the Far East.

The third category of animals in sport concerns those used by man for his pleasure: no hurt is intended and if anybody or anything is hurt or killed it is quite accidental. Horses are obviously the animals most concerned and they have been used for man's enjoyment from fairly soon after they were first domesticated; man found it not only useful but also pleasurable to ride the horse, and in Great Britain today it is the rare exception for the horse to be ridden purely for getting from one place to another. Even casual riding, or hacking as it is more generally called, is an expensive business and far beyond the reach of most. Horses are, of course, used in such blood sports as stag-hunting and fox-hunting, but for many people the only horses are those which race—whether on the flat or in steeplechases, or, drawing a light carriage, in trotting-races. Polo, which flourished particularly with the British Army in India in pre-war days, is now mostly played in those parts of the world where it is possible to keep ponies fairly cheaply. Perhaps the most recent development in sports concerning horses is the formation of pony clubs for children and young people, encouraging them to take a real interest in their charges and to organize outings, gymkhanas, etc.

Perhaps it is hardly correct to include greyhound-racing here, for it may be maintained that it is to coursing what the drag-hunt is to stag-hunting. It is an utterly artificial form of sport that is nothing but a means for extensive betting and there is little to commend it. Terrier-racing, on something the same pattern as greyhound-racing, is far more exciting.

Finally we come to a group of activities that would perhaps be more generally described as pastimes than as sports, though the distinction is not a very logical one and one section concerns what might be called Nature Study. There have been naturalists throughout the ages. Some of Solomon's writings make it clear that he was a competent naturalist: he knew the spider, the ant and the strange little rock hyrax or coney, and he noted how the gecko lived in houses. Jeremiah recorded a few hundred years later that the swallow and crane were migratory. St. Francis of Assisi was also a notable bird-lover. Most of the natural history writings of the Middle Ages, however, contain a large mixture of nonsense, certainly not based on observation, but towards the end of the 18th century Gilbert White set an entirely new standard with his *Natural History of Selborne*, a series of letters written between 1767 and 1787; this book has run into countless editions and is truly a classic. Nature students today are well off, particularly in these islands, for comprehensive volumes, beautifully illustrated, deal with every group of animals likely to interest; Witherby's *Handbook of British Birds*, for instance, is generally reckoned the finest of its kind in the world. Field Study Centres in various parts of the country offer wonderful chances of instruction in field work, and Natural History Societies exist to encourage the study of every aspect of nature. A recent list of such societies in Great Britain needed fifteen large pages in double columns and they varied, in name, from the British Snailwatching Society to the Malacological Society of London, though both seem to mean more or less the same thing.

Field natural history generally involves the collecting, and killing, of some specimens, especially in areas where the animals are little known, but as the work proceeds collecting becomes less necessary and observation more important, until in Great Britain, for instance, there is little excuse for collecting either birds or their eggs. Bird-lovers probably outnumber all other nature students put together; butterflies, moths and reptiles come next. Mammals are rather neglected, largely because they are generally more secretive and difficult to see. Bird photography has reached an astonishingly high level, but except for those reasonably classed as game, mammals have been relatively untouched in any country.

It may be agreed that entomology—or "bug-hunting", as it is often more popularly called—always involves killing, but the taking of short-lived insects, quite probably after they have laid their eggs, is a very different matter from killing the backboned animals. To assist amateurs there are even butterfly farms from which the rarer native species and exciting exotic forms can be obtained.

The keeping of fish in aquaria goes back a long way; the Romans certainly kept a variety of them, especially sea fish, and it seems that although the prime object was no doubt to have fresh food close at hand, the Romans also found these aquaria an interesting show. Fishponds continued to be popular and in the Middle Ages the monastic houses had their ponds, but these also were largely utilitarian. Fish culture reached a high development in the East and the work of the Chinese in producing special breeds of goldfish is considered in the discussion on domestication. The past two decades have seen the birth and growth of a new hobby, the keeping of tropical aquaria; many shops have been opened to sell appliances and stock, and most large centres have their Aquarist Societies where enthusiasts can get together. One of the great attractions of tropical aquaria is that they need to be kept at or near living-room temperature; in addition they are highly ornamental and without unpleasant smell.

Lastly, there is the keeping of pets, a custom found in almost all parts of the inhabited world. In this country the range of pets is most extensive and it includes almost every native bird and mammal that will live in captivity as well as many foreign kinds, and all the British amphibians and reptiles. In addition there are domestic animals, principally cats and dogs, which numerically and in value are far more important than all others together. This is right and proper, for the very first animal to be domesticated, the dog, very probably entered man's service as a companion and hanger-on in the hunting field.

2. ANIMALS AS PROVIDERS OF GOODS

From his earliest ages man has depended on animals for much of his living. The earliest inhabited sites excavated by archaeologists show a variety of animal remains: the bones from animals killed and eaten, often with marrow bones cracked and cleaned out: deer antlers and other bony parts made into weapons and ornaments. And it is virtually certain that in those days clothing consisted solely of animal skins. As civilization has progressed, so man's utilization of the animal kingdom has developed until it includes an astonishing variety of materials and uses from all groups of larger animals and very many of the lower forms. Animals have been domesticated in order better to organize the supply and these are treated in more detail in other chapters.

Food is certainly the most universal use to which animals are put and there can be few, if any, human beings who take no food derived from animals: even strict vegetarians generally allow themselves milk products and eggs. Fads and fancies enter largely into national food habits: in Great Britain, for instance, the edible snail is regarded almost with loathing by all but a few epicures—presumably because it is a snail. But so is a winkle and so is a whelk, both of them very popular dishes. In general, it is considered correct in Great Britain to eat shellfish from the sea, and the larger crustaceans from both salt and fresh water, but there is a very marked division into social classes, based no doubt in the first case on price. Cockles and mussels, tasty as they are, are generally spurned by the eaters of oysters and lobsters, or relegated to the realm of sauces and flavourings. In other parts of the world the lower creatures—the invertebrates—in the sea are much more widely used for food. There are, for instance, the squid and the octopus, and the sea cucumber, relation of the familiar sea urchins: the latter is sun-dried and sold as trepang or *bêche-de-mer*, which is a great oriental delicacy.

A still wider range of lower animals is enjoyed as food in less advanced countries, including a number of insects, a class very rarely eaten, far less enjoyed, by most Europeans. Locusts and flying termites (white ants) are taken in most parts of Africa, the latter more widely than the former: the Africans use ingenious methods of collecting them and then drop them into hot red palm oil—one method of deep frying! The grub stages of some insects are eagerly hunted: in the Gold Coast forest I found that the favourite was the palm weevil, whose grub, about $1\frac{1}{2}$ inches long and as thick as one's little finger, lived and grew fat in the heart of oil and raphia palms tapped for wine (the sap, slightly fermented, forms one of the staple West African drinks; it is full of yeast and probably does good by providing Vitamin B). My African staff often got a good haul of these grubs when we were out working in the forest, and once or twice I got Cook to prepare some for me: fried in a little butter and served on toast they tasted rather like and had something of the texture of soft herring roes, but I could not get out of my head what they looked like when we found them. In other parts of Africa, especially in Rhodesia, midges are sometimes eaten; they are so plentiful that they can be netted in thousands and compressed into a tasty form of cake.

The giant snail of West Africa is an important item in the diet of many tribes, and it has been claimed that in the Gold Coast it is, in fact, the most important single animal producing meat. It is a country where all kinds of meat are scarce, and the staple diet is a stodgy preparation rich in starches: such meat as can be got is made into a thick gravy to be taken with it. Smoked and dried snails make a particularly tasty stew, preferably with a liberal sprinkling of red pepper. These snails are truly giants, for they may reach the size of a man's fist. In the snail-hunting season whole families go out to their ancestral beats in the forest and camp there: the daily catch is brought to a central point where the snails are winkled out, skewered and smoked over an open wood fire. So many snails go to a standard skewer and so many skewers are fastened together to form a flat tray, the unit of sale, and there is a very complete and specialized vocabulary to cover all these points, just as there is in any ancient British craft or trade. Snail-hunting has been going on for many generations and it is so important that the beats are worked on a regular annual rotation, allowing the stocks to be maintained.

Birds, mammals and fish are so much the staple animal foods that only the more unusual need be mentioned. On the Gold Coast my work on animals included the preparation of many skins for the museums, and I had found in my years of collecting that almost everything left over after skinning operations was considered edible. In fact, I came to think that only two mammals of any size were not eaten—the hyena and a particularly foul-smelling little mongoose. After a while I moved to another area, where I found that this mongoose was very acceptable as food, but the hyena remained quite untouchable. Only two birds were refused—the hooded vulture and the pied crow. The vulture is the main scavenger in the markets and obviously unclean, but though also something of a scavenger, the crow seems to be refused for a very different reason: they say that many of its calls are nearly human and that human souls may, therefore, have taken up residence in it.

Bats were a favourite food in many areas where I worked, and caves where a kind of large fruit-eating bat lived were quite valuable to the village owning them. The dead bats were skewered in rows of ten or twelve and smoked: they could then be taken to the nearest market for sale. Their appearance put me off completely, quite apart from their very powerful smell, but the local people were not at all disturbed by such things and the native name of the common fruit bat was, literally translated, "It lies in the stew and smiles".

The Dagomba tribe of the Northern Territories was very fond of dog and even bred dogs for the table. A friend of mine was a Medical Officer at Tamale, the headquarters of the Dagomba people, and he once took up this question with a patient. He asked him how it was that his people liked to eat dog when plenty of antelopes, pig and fowl were available in

Picture Post Library

Feeding London pigeons in Trafalgar Square

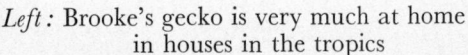

Left: Brooke's gecko is very much at home in houses in the tropics

G. S. Cansdale

A. R. Thompson

Right: Two house mice

Below: The swallow seldom nests away from buildings

Ronald Thompson

their country. The interpreter, a rather pedantic hospital assistant, translated the reply, "Sir, he says that the flesh of the dog far surpasseth in sweetness the flesh of all the other animals which you have mentioned." In most parts of Africa and India leopards seem to be in agreement with this patient, but I think few people could bear to try dogflesh, except perhaps in time of famine. In China the chow seems to have been developed especially for the table and its flesh was, and perhaps still is, highly prized.

In many countries, too, there is great prejudice against the smaller rodents, though hares and rabbits are eaten freely enough, but yet many smaller varieties are considered great delicacies in some places. The edible or fat dormouse (*Glis glis*) was bred for the table by the Romans, who even fed it on the most expensive items, including walnuts, to get the best flavoured flesh. In the Gold Coast practically every rodent—except that filthy parasite the black house rat—is eaten: at one end is the colossal crested porcupine, weighing three or four stone, and at the other end of those usually taken are the field mice, generally regarded as "perks" for the small boys. By the end of the harvest season these rodents are often fat and in fine condition, and there is no reason why they should not be good eating. The porcupines are always in demand and, even more so, their relation the giant cane rat or cutting grass, a solid creature rather on the lines of a coypu and weighing up to a dozen pounds. In my early service on the Gold Coast I found from time to time that an animal brought in by the hunters was described—when I asked if I could try it—as "No be white man's chop," i.e. it was not suitable for a European. Some of the things justified such a description, but I soon found that it was frequently used to get for themselves a tasty bit of some game such as a cane rat. After that I insisted on having my share of a cutting grass, especially if it was fairly young.

Squirrels are eaten in almost every country and the red squirrel was once sold by the thousand in the London markets. It is a pity that the grey squirrel has not been popularized as a dish, for that might increase the price on its head and thus keep its numbers down to a more healthy level. The guinea-pig is firmly established in the civilized world as a laboratory animal and pet, but when it was first domesticated in South America it was used for sacrificial purposes and for food. It is still bred for eating in West Africa and, presumably, in Peru.

In a recent publication by the Wine and Food Society, M. André Simon has listed well over 100 mammals which, as he notes, "are commonly eaten or have been occasionally eaten by man": most of them no doubt come in the latter rather than the former class. Mr. L. R. Brightwell, best known for his animal drawings, seems to have experimented widely in this gastronomical field and he is quoted widely on the palatability of many strange meats. He found, for instance, that the badger's flesh is rather rich and porky, while bear of any kind is excellent though requiring slow cooking. M. Simon does not content himself with listing these animals, but in many cases also gives recipes which certainly sound appetizing enough.

Apes and monkeys are considered taboo in many parts, presumably because they are regarded as too closely similar to man, thus perhaps being subject to the adage "Dog does not eat dog". In Africa monkeys are generally avoided by tribal groups within the Moslem orbit and also probably by those having easy access to supplies of more orthodox and larger game, but they are greedily hunted by many forest tribes, and where this happens they seldom, if ever, become a serious pest. When out in the forest I tried monkey stew once or twice and would probably have enjoyed it had I not known what it was.

Many of the old tribal controls are now disappearing, but in the more undeveloped parts of the Gold Coast there are still very rigid food taboos, inherited through the male line. In a land suffering from an acute shortage of animal protein this seems a rather senseless deprivation, especially when a man might be forbidden such staple meats as chicken, beef or mutton. The food prohibitions imposed on the children of Israel were not just ceremonial taboos; they were mostly based on sound hygiene in a hot country. The ruminants—antelopes, sheep and cattle—were allowed, but the pig, largely a scavenger and the potential carrier of disease, was forbidden. Among the birds the chief prohibitions were the flesh and carrion feeders. It is interesting to note that four different sorts of locusts were specifically allowed as food and, some centuries later, they were an important part of John the Baptist's diet.

Birds generally are the subject of fewer prejudices, but in normal times we are much less catholic than our forebears were. We would hardly relish a dish of roast swan, peacock or crane—or is it that we have just lacked the opportunity of tasting them? Perhaps it is as well for our song-birds that "Four-and-twenty blackbirds baked in a pie" are no longer reckoned such a "dainty dish to set before a king". But in times of meat shortage we become less inhibited and during the recent war size and availability were often the chief, perhaps the only, factors that decided whether a bird went to market. A bird's feet can be too revealing and in very flagrant cases of deception and misnaming they were removed; even beaks have been trimmed to a different shape. Until well into this century certain sea-birds were taken in thousands by the islanders around the Scottish coasts; the gannet, fulmar and puffin, all of them colonial nesters, were the principal kinds used. They were cleaned, opened out and dried in sun and air to form the main winter meat stores for these isolated communities. Richard Kearton, photographing the sea-birds in the 1890s, found the St. Kildans at work on the cliffs, catching the birds by centuries-old methods. Puffins were caught in very great numbers, partly in long lines of snares and partly by the use of a horsehair noose on a 13-foot rod: in this latter way up to 600 were bagged in a day. It was estimated that something like 90,000 puffins a year were killed around St. Kilda alone. The fulmar was even more valuable to the islanders, for it provided lamp-oil and feathers as well as meat, though to other folks the smell made the fulmar useless.

Birds' eggs are enjoyed almost universally; in Great Britain the preference is for them new-laid, though the marketing methods developed in the mid-20th century make it more advisable to have the weekly egg opened before it comes on the table. I always adopted this method in West Africa with eggs brought as presents while I was on tour; it was much better for Cook to have the surprise. Perhaps it was not really a surprise, for he seemed to expect a proportion of these gift eggs to "catch piccan for inside", i.e. they had been sat on for some time. Tastes vary, and the Chinese apparently enjoy their eggs best when they are very old. The oddest bird products to be eaten are their nests; certain oriental swifts make small cliffside nests out of their dried saliva and these are highly esteemed by the Chinese for soup-making.

The prejudice against reptiles as food is fairly general and I confess to it myself! A thick white steak from the tail of a large monitor lizard or one of the fish-eating crocodiles looked most appetizing but I never brought myself to try it. One tribe in the Northern Territories of the Gold Coast specialized in crocodile catching and they were called in periodically to reduce the numbers of crocodiles in the water-supply reservoirs. They could catch them in their nets only when the water was fairly low and so they kept them trussed up alive until they were wanted for eating.

With snakes it is partly a question of size. Most kinds are far too small to be worth bothering with, but a large python or a five-foot puff adder (weighing perhaps 15 lb.) can yield thick white steaks, looking something like cod. During my collecting activities I found that although not everybody would eat snake, there was enough demand, especially among the itinerant labourers, to fix a more or less regular market value; to get my snakes alive, therefore, I normally had to offer something like twice the standard meat price as well as sending out to take delivery. In the U.S.A. rattlesnake is considered a delicacy; it is most acceptable out of a can—perhaps because it is less recognizable in that form.

Turtles and tortoises are reptiles just as truly as are snakes, lizards and crocodiles, but to most folk they probably seem in a different class altogether. Quite probably the alderman enjoying his turtle soup at a civic banquet would like it less if told that he was eating reptile. Terrapins—a group of freshwater tortoises—are farmed in parts of the U.S.A. and sold for their meat. The land tortoises are also taken, though they are vegetarian in contrast to the carnivorous sea terrapins and turtles, and they are farmed in parts of Africa and in Japan. In the days before refrigeration, ships used to call at the Seychelles Islands and pick up a number of the giant tortoises formerly so common there. They stayed alive on board for weeks, or even months, and they were a very valuable source of fresh meat. Their full story is told in another chapter.

As a class, amphibians are used less than any other vertebrate animal; in fact the French, with their fondness for frogs' legs, seem almost unique. Some Central African tribes also enjoy frogs, even preferring them alive. Prejudice may enter into this abstinence—it certainly does as far as the British are concerned—but it is probably due mostly to the small average size of most amphibians and the unpalatability of many of them.

With fish the position is rather simpler; there are some prejudices due to colour, appearance, shape or just plain old wives' tales, but it is mostly a question of being palatable. Size matters not at all and a dish of fried whitebait is as acceptable as a cut from an immense halibut. The average freshwater fish in Great Britain, apart from the trout and salmon, is left uneaten because of taste; even the keenest angler must admit that roach or chub is not really worth cooking. The eel, however, is something of an exception; it spends most of its life in fresh water, having arrived there as an elver weighing a few grains, but it eats as sweet as any fish. In Great Britain it is greatly appreciated, mostly in jellied forms, by certain sections of the people only, and its demand is such that a well-known chalk stream fishery normally gets as much income from the eels trapped, as they come downstream to start their breeding journey to the mid-Atlantic, as it gets from letting its world-famous trout beats. On the Continent there is no class prejudice, for smoked eel is rightly considered an ingredient of the very choicest *hors-d'œuvre*. The Romans also knew what was good; they established fish-ponds where they could keep and fatten their favourite fish and they preferred such varieties as mullet, sole and turbot, but most of all they went in for Roman eels. They could not have been worried by the appearance of their food when alive, for the Roman or Moray eel has the wickedest-looking head of all sea fish, excluding perhaps some of the deep-sea monsters.

In the Middle Ages carp-farming became the vogue in Europe, mostly because fish was allowed on otherwise meatless days; in the days before fast transport and refrigeration, the inland areas could enjoy no fresh sea-fish, nor had methods of preservation been properly developed. Fish-farming has been developed to a high degree in parts of the East and, properly managed, a fish farm is a most efficient way of producing animal protein.

Fats come from two sources—from plants and from animals. Animal fat is now made into a great range of products, both edible and industrial, but it has had a variety of uses since very ancient times. Among the Hebrews it had considerable significance in the complicated sacrificial rituals. Fat has also been an important ingredient in medicines and ointments. In ancient Egypt, for instance, the Hearst papyrus records that one remedy for making the hair grow was compounded from the fats of the gazelle, serpent, crocodile and hippo. Another Egyptian recipe called for a mixture of lion, hippo, crocodile, cat, serpent and goat fats, but its purpose is not suggested. Some animal fats are still in demand for making country remedies: in West Africa the large fat bodies lining the abdomen of giant vipers were used for concocting a potion.

It is a little difficult to classify the very diverse animal products other than meat that are used by man. By far the greatest volume of such materials, and the greatest in value also, is derived from domestic animals, many of which have been brought into the service of man for that very purpose: other animals were first domesticated for some different reason and then developed along other lines also. The part played by these animals is discussed in another chapter, together with their important products. In contrast to domestic stock, wild animals yield comparatively little of value while still alive; it is their dead bodies which are used. Eggs, strictly speaking, are not an exception to this rule, for an egg is alive and is utilized only by being killed in or before eating. Eggshells also have some uses, though most kinds are useless except for ornament. Ostrich eggs come in a different class and they were used as water containers in pre-pottery times some 5000 B.C. in the Libyan Desert. Beads made from ostrich eggshells have been found associated with the same culture.

Guano, a product of excretion, seems the only important exception to the rule. Guano is the dried or semi-fossilized droppings of countless thousands of several kinds of sea-birds, something like the cormorant, that live on the islands off the south-west coast of America. These particular islands are situated close to the Humboldt current, which is immensely rich in all sorts of fish and therefore provides unlimited feeding for the birds. Guano can be formed only in rather special climatic conditions; in particular, there must be virtually no rain or the deposits can never accumulate. The absence of these climatic conditions on all islands near the British coast means that guano can never be formed, however plentiful the sea-birds and copious their droppings. In the guano islands the deposits are forming all the time, but most of the material removed had been produced by generations of birds long ago.

Guano is made only by the gregarious animals, and apart from these birds bats seem to be the only creatures whose droppings are utilized. Even though they are bound to live in areas with at least a moderate rainfall, the colonial insect-eating bats of the tropics normally roost under cover—in caves, hollow trees and houses. The south-eastern part of the Gold Coast is inhabited by the Ewe people, probably the most enterprising of all West African folk and certainly the most able farmers. In that part too there are enormous populations of gregarious bats whose only available sleeping places in the densely populated coastal zone are the houses: these roosts are cleared out regularly and the dry powdery guano, remains of millions of insects, has a definite market price. Other tribes are now beginning to learn the value of animal manuring, but when I was working among these people in the 1930s they were more or less unique in their appreciation of it. They even made arrangements with the cattle-owning nomads to tether their cattle at night in the coconut plantations. Bat guano is probably used in various other parts of the tropics, though not by any means all native people would know what to do with it. The Dyaks, people of North Borneo, have access to large supplies in caves which the bats share with the swifts that make edible nests. Dried droppings, particularly from camels and cows, form a most important domestic fuel in parts of India and in desert regions, often to the detriment of the farms, where they should be used as manure.

The venom of some snakes finds a use in general medicine as well as in the preparation of anti-venins used to treat persons suffering from snake-bite. Although a certain volume can be obtained from the venom of dead snakes, the standard way of getting it is to "milk" the living snake, a job calling for great care and coolness. The drops of liquid venom are then desiccated and the crystals can be stored without deteriorating.

Finally there are the many non-edible materials made from parts of the dead animal. Skins were used very early in the history of mankind, and among primitive tribes today they are, as in the beginning, their only body covering. It is a far cry from that to the natural mink or sea-otter coat costing several thousand pounds. A list of animals whose skins are taken for use as fur would cover all the main groups: monkeys and lemurs in the primates; squirrel, beaver and nutria in the rodents; leopard, ermine and sable in the land carnivores and seal in their sea cousins; wallaby and opossum in the marsupials and pony and dik-dik in the ungulates. The sheep and rabbit are character artists in the hands of the furrier and are groomed to masquerade as a host of much more aristocratic cousins.

The craft of leather-making was invented soon after the first big surge of domestication of animals, and it has probably always been based on the use of domestic animal hides. Leather was known in the Tasian and Badarian periods of pre-dynastic Egypt in the fifth millennium B.C., and the ancient Egyptians achieved considerable proficiency in its preparation and use. Over in the Middle East, leather-work had already reached a high state of development by the Third Dynasty at Ur, 2050–1950 B.C. The tanners and leather workers were well organized as trade groups: dyed gazelle skins were used, sometimes six or seven skins to one article, and leather was graded for quality. In both Mesopotamia and Egypt dyed leather was made into sandals, bottles, bags and purses, while parchment, also prepared from animal skins, was used for drumheads and the sounding-boxes on musical instruments. Some wild skins can be used, but cow leather remains supreme for hard wear and goatskin for fine work, while the classical material for men's gloves is hogskin or deerskin. More exotic and ornamental leathers have become popular in fairly recent times, including sharkskin, snakeskin (mostly from African pythons and an Indonesian water snake), lizard and crocodile (only the belly skin). The monitor lizard is the principal sufferer among lizards, and in some parts of Africa it has been hunted so drastically for

Zoological Society of London

The London Zoo in early Victorian times

Elephant rides are always popular at the Zoo

Sport and General

Picture Post Library

Feeding sacred ibises in Ancient Egypt

its skin that even the crocodiles, whose eggs are its favourite food in some areas, have increased in numbers as a result. The skins of a few birds, including the ostrich, are used from time to time, so that the amphibians are the only group of backboned animals not to be utilized in this way.

Feathers are used for various jobs, both utilitarian and ornamental, and it is rather obvious they can come only from birds. The incomparable soft feathers which the eider duck uses to cradle her eggs have given the name "eiderdown", though it is probably only a small proportion of eiderdowns sold today that are actually filled with what they claim. In one or two parts of Iceland the gathering of eider down is a carefully organized local industry: the eiders nest in fairly large colonies, encouraged by the Icelanders, who make little nooks especially to attract them. The down is collected after the ducklings have hatched and safely left the nest. Feathers have been used for stuffing mattresses since early days, and Chaucer talks about using pure white doves' feathers for the purpose. The term "feather-bed" has also long been used in a figurative sense: the first recorded reference, in fact, is to a "feather-bed captain" well over 250 years ago.

Quill pens still have their devotees, and the Zoo is regularly asked for large quills, preferably from condors or eagles, for making fishing-floats, but the most exotic request in this line received for some time was for some peacock quills with which to make a plectrum for an 'ud. This, we were told, was an Eastern musical instrument used in the B.B.C.'s Arabic service. When the peacocks had at last shed their tail feathers (actually their upper tail coverts) they were sent along, only to prove not rigid enough for the job, and so we had to try again.

Larger feathers are mostly wanted for ornament, especially on ladies' hats. Most of these feathers have been moulted naturally and many others come from domestic birds, so that their wearing does little harm, but we can only deplore strongly the use of large parts of the plumage of brightly coloured birds. At one time birds of paradise suffered severely at the hands of plumage-hunters and even humming-birds have been used for hat trimming. Fortunately for the birds, fashions have changed and brought more relief than protective legislation. Feathers are also in demand for making salmon flies: these lures imitate no known insect but yet have a strange and sometimes fatal attraction for the salmon. Various colourful feathers are used, but the one in greatest demand is the hackle of the Sonnerat's jungle fowl, which seems to add the killing touch to Thunder-and-lightning, Silver Grey, Jock Scott and several other West-country favourites. The much more demure trout flies, on the other hand, though also made mostly of feathers, generally imitate closely one or another stage of a number of aquatic insects.

Sheep's wool is by far the most valuable animal hair used by man, but this is today a product of domestication; so are the pigs' bristles used for hairbrushes and the horsehair which still has sundry domestic uses, though it has long been out of date as a fishing-line. But hairs of wild animals have various uses; the badger, for instance, still provides the best material for shaving brushes. In ancient Egypt, long coarse hairs, probably from the tails of oxen, were used for making bracelets or for threading beads. Similar hairs went into the making of fly switches, but in Roman times the most fashionable ladies used yaks' tails. One might hardly expect the intestines of animals to produce anything in the fibre class, yet the catgut used for tennis racquets is prepared from the intestines of sheep—preferably thin, undernourished sheep. Cats may have been the source at one time but they are never used for this purpose nowadays. The oldest known use of gut seems to have been in the Badarian period of Egypt, dated about 5000 B.C.

Other animal parts gave man some of his early tools and tool handles. Teeth of all shapes and sizes, deer antlers and bones, these were all invaluable to our forebears. The antlers of the large deer so typical of northern temperate zones proved most useful, for they were shed annually and were to be had just for the picking up; the earliest known sites of human habitation in Great Britain are rich in deer antlers and in tools made from them. Bone was used extensively by the people who made the earliest permanent human settlements yet investigated: the village of Jarno in the Middle East is dated at approximately 4500 B.C., and its excavation has yielded hafts, awls, needles, beads and spoons made from bone. At about the same time other peoples in Egypt were also using fish bones, the vertebrae as beads and the thin bones as needles.

Metals have replaced bone for most cutting jobs today and antlers are now used for ornamental work and handles. Teeth also are now used chiefly for ornament, whether it be sharks' teeth to form a South Sea Island necklace, or ivory, probably the most perfect medium for the fine craftsman. True ivory comes from elephants' tusks and these, with hippo teeth, have been used for carving since Neolithic times.

The hollow horns of the cattle tribe were early used for making cups, bracelets and other ornaments: they also provided one of man's earliest wind instruments, perhaps the very first, and the name is still found in one of the symphony orchestra's standard brass instruments—the French horn. The horn was also the traditional powder flask, and it is so today among some African tribes who still use muzzle-loading guns. Many strange parts of animals are used for medicinal purposes, but perhaps the most perverted taste of all is seen in the demand for rhinoceros horn as an aphrodisiac in the Far East. It is worth far more than its weight in gold and thus puts a frightful price on the head of all the rhinos, whose position, in any case, is by no means secure.

Although scent production is largely the prerogative of the plant world, or was until the chemist took over quite a bit of it, the animal kingdom produces several materials still needed by the manufacturer. Ambergris, produced by the sperm whale and found largely by chance, has the property of scent-holding; civet secretion, from the live animal, and musk, in pods obtained only by killing the Asiatic musk deer, are also still used extensively in perfume-making.

The lower animals give mankind products varying from dyestuffs to fibres. The Tyrian purple dye came from a shell, while cochineal, the red dye used mostly for cookery, is produced by an insect belonging to the same group as the woolly aphis and green-fly. Cowrie shells have long served as money and ornaments; other kinds of marine and freshwater shells are made into necklaces, while some of the larger ones are used as receptacles. Mother of pearl, obtained from the inner lining of various oyster-like shells, was being used for making bracelets, buttons, pendants and rings well over 6,000 years ago: at the same time tortoiseshell from both land and marine tortoises and turtles was being made into combs, bracelets and parts of musical instruments. Shells in bulk, on the other hand, may be a valuable source of lime in countries without supplies of lime-bearing rock. Spiders contribute the finest of web for making the crosswires in various optical instruments. Fishing-casts were formerly made of a fine translucent gut prepared from the intestines of silkworms: some of the heaviest grades are still best made of gut, but for most types artificial fibres such as nylon have proved just as efficient, as well as being far cheaper and more durable. This catalogue could be prolonged endlessly but this is perhaps enough to give some idea of the vast range of direct uses to which man puts products of the animal world.

3. HANGERS-ON

IN most kinds of relationship between animals and man discussed so far the initiative has normally been taken by man, though in some cases, especially when it comes to introducing animals to new areas, the animals have a habit of taking matters into their own hands again. But there is also a large and assorted group of animals which have been attracted to the neighbourhood of man—some as parasites, some as hangers-on and many just as associates. In this connection the word "parasite" is used in a broad sense and not with the narrower zoological meaning, for a body parasite, such as a flea, has exactly the same relation to man as it has to any other animal —and very few animals, whatever their size, are without parasites.

It is man's activities that have made most of these associations possible: his buildings provide nesting sites and his cities new realms for enterprising animals to occupy, while his habits of growing and storing masses of foodstuffs and other commodities encourage the parasites. Cases of animal association in the wild are not unduly rare—such, for instance, as the habit of some weaver birds to make their nests underneath a large eagles' eyrie, and the attendance of tick-birds, or ox-peckers, on cattle—but these relationships are fairly simple and any one animal seldom has such associations with more than one or two others. When discussing the problem of domestication it was noted that the ants and termites were the only animals, other than man, that went in for employing other animals: it is very much the same as regards hangers-on. The great assortment of creatures that live with ants—several hundred are recorded for British ants alone—have been classified as true guests, indifferently tolerated lodgers and hostile persecuted lodgers, in addition to the true parasites. One could perhaps use almost the same groupings for the animals associating with man.

The rats and mice are economically the most important of these and we would have no difficulty in classifying them as "hostile persecuted lodgers". This is the chapter where they really belong, even though they have already had brief mention as domesticated laboratory animals and as hitch-hiking vermin. The house mouse is often assumed to be the oldest human associate, but rats and mice are often confused, and one cannot be quite certain what is meant in some of the ancient records; this confusion is found in West Africa today, where almost every rodent from the mouse to the brown rat is known as "mouse", and the word "rat" is reserved for an edible giant rodent, correctly called the Gambian pouched rat. The house mouse today seems completely dependent on man for its food and it spends most of its time actually inside buildings, though it is prepared to go to the cornfields for the summer. It came originally from Asia and it was probably a pest in Egypt several thousands of years B.C.: it spread over most of Europe in early times and is now found virtually all over the world, but there is little to show just when it came to this country. It had, however, been on St. Kilda long enough to develop a distinct form, though it is presumed that when the island was evacuated in 1930 the St. Kilda house mouse was soon lost for ever. Even the most ardent conservationist is not likely to grieve over such a lost sub-species. The house mouse is a pest everywhere; if it consumes less food than the rats, this is only because of its smaller size, but it also destroys much that it cannot eat.

More is known about the rats. It is generally agreed that the house rat, or black rat, came to Europe in the early days of this era and was already widely distributed there in the 13th century, but it has also been found in fossil form dating from prehistoric times, associated with the lake dwellings of Switzerland and the Bronze Age in Hungary. Presumably this early

ASSOCIATIONS BETWEEN ANIMALS AND MAN

form had died out and Europe was then colonized from Asia, the original home of the house rat. It travelled in the early ocean-going ships and is known to have arrived in South America by 1540: it is very much at home in warm countries.

There is no doubt that the more enterprising and aggressive brown rat (misleadingly called the Norway rat) is a much more recent arrival: it is generally thought to have reached Europe in the 18th century, partly assisted by man, partly as a result of mass migrations. It is authentically recorded that large troops of brown rats crossed the Volga from Central Asia about 1727, making in a westerly direction. It probably arrived in England about 1730 and its presence in North America was first confirmed in 1775. The brown rat is a heavier and much more successful animal, and it has largely turned out the older form, but it is more a rat of the lower levels, including sewers and drains, while the house rat is a better climber and survives in places inaccessible to the other; it is also the regular ship rat. "Brown" and "black" are unsuitable descriptive names, for in both kinds the colours vary; the brown rat is distinctly bigger and has a tail shorter than the head and body: in the house rat the reverse is the case.

It is quite impossible to estimate their numbers in any country, and the total amount of damage done by rats is also incalculable. A fairly recent estimate for the U.S.A. was 200,000,000 dollars per annum, and in many other countries the figures are just as appalling. Another major indictment is the carrying of bubonic plague—the true name for the Black Death and the Plague of London in 1665. This killing disease goes back to antiquity and it may well have been the cause of the fatal tumours from which the Philistines suffered in 1 Samuel v and vi: this outbreak certainly coincided with a plague of rats.

These rats and mice have shown themselves to be extraordinarily versatile and able to make themselves at home on virtually any type of food in every climate from the rain forest on the Equator to the whaling station of South Georgia in the sub-Antarctic. They can live even in the refrigerated holds of meat ships, where they grow a long protective coat. No other rodents have proved anything like so succesful as these: in India one or two of the bandicoot rats, and in West Africa at least two kinds of rat, one large and one small, have taken to a town life, though as individuals rather than as whole species, but they are not seafarers and are never likely to be more than local nuisances.

In fairness to the rodents it should perhaps be pointed out that over 550 kinds of true rats are known, native to almost all parts of the world except America, as well as very many others that can best be described as rats and mice. Out of this great number only *three* species have become parasites on the human race: most of the rest are well-behaved little animals that prefer to keep themselves to themselves.

No other mammals seem to have become firmly attached to the human race: in fact, the only other higher animal at all comparable with the rats is the house-sparrow, and one or other form of this bird seems to be found in the neighbourhood of man in most parts of the Old World. Here also the association is of long standing, for the Palestine sparrow is mentioned as a familiar bird in both Psalms and Gospels. The British house-sparrow seems to have become more dependent on man than most kinds and Nicholson, in his recent book *Birds and Men*, even suggests that "the disappearance of the human inhabitants from Britain might well lead to the utter extinction of the house-sparrow". It is essentially a town bird and although flocks may go to the fields for part of the summer they cannot live there permanently. Nicholson points out that because of its town habits the sparrow's numbers in Great Britain as a whole are often grossly exaggerated and it is, in fact, much less numerous than the chaffinch, blackbird and several others. Even so, it is more plentiful than most people would like it; the general opinion is that it does far more harm than good, for it eats cereals in one form or another almost the whole year through. It should, in fact, be classed as a hostile persecuted lodger, even though many people encourage it and most tolerate it. The decrease in horse-drawn traffic was a sore blow to the sparrow because the horse droppings, with their content of undigested oats and other seeds, formed an important part of its town diet, and its numbers have consequently dropped; but perhaps there has been only a redistribution, for the sparrow is most adaptable.

One of the biggest complaints against the sparrow is that

it turns out the house-martin, and in some places there has been a very serious decrease in the numbers of this attractive and useful insect-eating bird: as soon as the martin's nest is completed the sparrow takes over, and when the martin can at last get settled it is probably too late in the season to rear the young. Sparrows also do a great deal of apparently unnecessary damage to cultivated flowers: they also delight in pulling up seedlings and stripping buds from all kinds of fruit trees and bushes. They are quick to learn new and unpleasant habits. When I moved into my house at the Zoo in 1948 I found a useful small vegetable garden in which I raised some crops, including a lot of lettuce. The following year I again had beds of lettuce and all went well until, in a dry spell, the sparrows began to peck the lettuce leaves: I put out more water for them, right alongside the lettuces, but they preferred the leaves. Since then I have been quite unable to grow lettuce unprotected: the wretched sparrows even go into the glass frame for early lettuces if the top light is opened more than two inches. Taken by misguided colonists to new lands, sparrows soon became a pest there too. In North America they were already being condemned as an unmitigated nuisance in the 1860s, and in New Zealand they now do immense damage to the corn crops.

In many people's minds town pigeons—particularly the London pigeons—are very much associated with the sparrows, though they can reasonably be classed as true guests; the experts, however, scarcely condescend to admit that they exist, and the *Handbook of British Birds* makes only a few disparaging references to them as "feral dove-cote pigeons". It is true that these pigeons are the descendants of tame pigeons which, in turn, were derived from the wild rock dove, but London had its pigeons at least as far back as 1385: although the stock is given fresh blood from time to time by new escapes, the London pigeons have, in fact, a long ancestry, and they must almost be regarded as a revised edition of the rock pigeon, in which the habits of cliff-nesting are modified to make use of buildings. The town pigeon is even more closely and permanently tied to the immediate neighbourhood of town-dwelling man than the sparrow, which may take a seasonal holiday in the country.

Many other kinds of birds also frequent the neighbourhood of towns or villages to a greater or lesser extent. Some are attracted by feeding or positive protection, while others find that gardens and farms provide ideal feeding or breeding grounds. These would include most of the finches, buntings and thrushes, some of the tits, and such well-known favourites as the robin, wren and hedge-sparrow. This aspect is dealt with by E. M. Nicholson in great detail in his book referred to above.

The jackdaw and the starling are two other town birds which also demand some notice, though at best they are barely tolerated. The jackdaw—a sociable member of the crow family—is, in the wild, a crag- or tree-nester, but buildings old and new, especially church towers, provide excellent nesting sites. It is a most adaptable bird and well fitted to make use of new territories and foods, as well as new nesting sites and materials as they become available. Nobody who works or lives in London can fail to notice the roosting starlings, but, extraordinarily enough, it does not seem to be known precisely when they started doing this. They are first mentioned as roosting in increasing numbers on the islands of some of the parks towards the end of last century, but these roosts used to be vacated in the autumn. It is suggested that London buildings were first used as roosts during the First World War, and by 1922 many well-known buildings were being occupied in autumn and winter by large flocks. Most of the birds were first thought to be migrants from the Continent, but Nicholson proved that they actually came from no farther afield than the suburbs. These birds are a considerable nuisance: they foul the buildings and are a constant menace to pedestrians on the pavements. At times their chatter is annoying, and they have the embarrassing habit of sitting on clock hands in the horizontal position and making the clock slow: no successful method of starling-scaring has yet been devised. Precisely the same town-roosting habit has developed in many other British cities and, as already described earlier, in American cities as well.

In recent years communal roosting in urban surroundings from autumn to spring has been developed by the pied wagtails in the neighbourhood of Dublin. They have chosen a short row of trees not more than 30 or 40 feet high in one of the streets that is busy until late every night, and the long-tailed black-and-white birds stand out very clearly in the leafless trees. Numbers grow each winter and more trees are being occupied, presumably by birds which gather from a wide radius, for they are not gregarious by day.

Perhaps we should mention here the black redstart, one of the latest additions to the list of regular British breeding birds, for, like the wagtail, it is a most welcome guest. The first few records were in southern England in the early 1920s, but regular nesting began in 1926 on the old Empire Exhibition ground at Wembley, where three pairs nested in an almost deserted factory bulding. In parts of Switzerland, Austria and southern Germany this charming little bird almost takes the place of the robin in England: it generally sings from the house-tops and nests on ledges of buildings, and is, in short, a bird that deliberately chooses human habitations as its breeding territory. When the first few colonists came to Great Britain they did not change their habits and Fitter, in *London's Natural History*, gives the full story of the occupation, a story featuring many famous London buildings. He points out that the blitz is wrongly considered responsible for the black redstart's move into London: the move had, in fact, started well before the war, but the bombing made many suitable nesting places, and the open sites, colonized by all manner of plants, served as hunting grounds.

In London, gulls are as familiar in winter as are the roosting starlings and they, too, are comparatively recent arrivals, for the habit of spending winter in and around London started in about the 'nineties, probably as a result of gales driving large numbers of them up the Thames. This new way of living was first worked out by the black-headed gull (without its dark hood in winter but recognizable by red beak and feet), but it has since been copied by several other kinds, including the herring gull, whose mewing call is used by the B.B.C. to suggest a sea-cliff setting. In winter, gulls are largely scavengers, and they find a living on the river, at rubbish tips and on sewage farms, but they do not object to charity and they compete with pigeons for scraps of bread thrown by visitors to the parks.

Towns in most parts of the world have their bird attendants.

In the warmer zones much of the scavenging is still done by the vultures, which sit around the markets as impudently as pigeons around a London square. In West Africa, at any rate, the hooded vulture gets little in the way of meat and has mostly to be content with vegetable matter and refuse. The pied crow is also very much of a town bird, and is seldom found far from the vicinity of man, but it is as fond of stealing chickens as of scavenging. While the tawny owl is the typical London owl and can be heard regularly in all the parks, it is the barn owl that has colonized many African towns, using the roofs of ordinary buildings as nesting sites and doing good by its rat- and mouse-catching. Another bird of prey has learnt to take advantage of modern traffic. In parts of Scotland rabbits are very common and many are run over, especially at night: buzzards have formed the habit of frequenting the moorland roads to take over these remains.

Man's activities have obviously created many situations of which birds can take advantage, but there are two groups of birds which now make general use of human surroundings as nesting sites and are very welcome guests. These are the swifts and the swallows, rather similar in general shape and habits, but actually belonging to quite different families. Formerly cliff-nesters for the most part, many species are now found nesting in close association with man. In the Gold Coast we had about a dozen kinds of swallows and martins, and they could be found nesting in old rest-camps and modern houses in European style as well as in road and railway culverts. At one of my stations a little crag-martin nested regularly under the eaves of my office and, between flights, often rested on ledges around my bungalow farther up the hill. I rather took the little bird for granted and kept only brief notes on its comings and goings and its nesting activities, but when the appropriate volume of the standard work on West African birds appeared some years later I found that mine was almost the only recent report about it: early observers had known it literally as a crag-martin with a very limited distribution, but when man put up stone and cement buildings it was able to extend its range. A river-martin, very smartly clad in white and electric blue, even nested in the pontoons of busy ferries on some of the big forest rivers, though the disturbance

Mummies of cats from the XVIIIth Dynasty of Egypt in the British Museum

W. F. Mansell

A white horse on a hillside near Thirsk, Yorkshire

Picture Post Library

The arms of Barclays Bank   The Great White Horse, at Ipswich

(*Below*) The arms of Martins Bank and the arms of the Royal Veterinary College

Picture Post Library

generally seemed to prove too much and the nests were often deserted. Even the palm swifts seemed to prefer nesting in the palm trees planted around houses and as town avenues. In Great Britain this association is just as intimate, with the swift and house-martins under our eaves and the swallows in our barns and outbuildings: even the sand-martin today is more frequently found nesting in man-made banks such as quarries and railway cuttings than in natural cliffs.

One or two members of the largely tropical family of weaver birds often choose the neighbourhood of villages for their nesting sites. One West African species, probably the commonest kind in the whole area, is generally known as the village weaver, and one commonly sees trees smothered with nests in the middle of clusters of houses, the birds going on with their activities without worrying about people working only a few yards away. These weavers get their nesting materials by stripping palm leaves, with the result that palms standing close around the village are often a sorry sight. But these weavers do also nest away from human habitation, sometimes a considerable distance away. Several much smaller weavers, the indigo finch and one of the fire-finches especially, and to some extent the mannikins, are also very typical breeding birds of the villages in parts of West Africa.

The only group of mammals which can be at all compared to the swifts and swallows are the insect-eating bats, their opposite numbers in food habits. In temperate regions bats have two main types of natural roosting or hibernating places —hollow trees and caves: buildings can replace either of these and it is perhaps not surprising that eleven of the twelve British bats are known to use buildings. In tropical zones the space between the ceiling and roof is most attractive to bats: it is often necessary to practise bat-proofing, for if bats really take possession they may foul the premises so badly that they have to be abandoned. In such cases they cease to be guests welcome for their insect-catching habits and become hostile persecuted lodgers. Road culverts in West Africa are also favourite roosting places: in several long series inspected in the Gold Coast I found that on the average every second culvert was tenanted. It seems possible that both bats and birds are first of all attracted by the easier hunting over clearings and around buildings and farms; after that it is perhaps natural for them to look for suitable nesting places on the spot.

Great Britain does not seem a particularly attractive region for reptiles and amphibians: there are only about half a dozen species of each and, except perhaps for frogs, they are seldom abundant, but in moist tropical areas both groups are immensely more common and in many parts there are several varieties of lizards that take over man-made habitats. In West Africa the large Agama lizards swarm on the roadsides, in the compounds and on houses, doing great good in their constant search for the insect food which is so much more abundant in the clearings. At night the little geckos come into their own and gather a rich harvest of insects attracted to the light, while a few large toads are likely to come on to the verandah to scoop up any that fall to the ground. Even the cobra is a sort of hanger-on, for it finds its living easiest on the edge of human habitations where rats, lizards and toads are plentiful. The classes into which these animals are placed will vary very much with personal taste!

When we turn to hangers-on among the lower animals we find that, like those that have been chance introductions, they are a thoroughly bad lot. Several flies are closely associated with man, and the house-fly is never found far away from him: it has followed man everywhere and is almost as much at home in the Arctic Circle as on the Equator. These flies are always a filthy nuisance, the spreaders of many diseases, and it is well that they are susceptible to attack by D.D.T. and other new insecticides.

Cockroaches are, if anything, more loathsome than flies, and two kinds are firmly established in Great Britain as house pests. The common cockroach is found in almost every town and village; it arrived by sea during the 16th century and has now been following man around for so long that its country of origin is quite unknown. The so-called "German" cockroach came to Great Britain much later—less than a century ago—and did not come from Germany! There, in fact, it is called the Russian cockroach. Both of these unpleasant insects are confirmed house-dwellers and need warmth and dampness: these conditions are found in most old kitchens, as well as an adequate food supply. They are difficult to eradicate completely but can be kept at a low level by constant attack. To most country folk they are "black beetles", but are neither black nor beetles! Two much larger kinds are found in permanently warm places, including the heated houses in zoological gardens, and these are always in demand for dissection by zoology classes. The house cricket, belonging to the same group of insects, has rather similar habits, but never seems quite so unpleasant.

Although some clothes moths are rather catholic in taste they now seem, as a group, fairly definitely committed to a world-wide indoor existence. The damage they do was noted long ago and Job, writing probably about 400 B.C., but possibly very much earlier, referred to a moth-eaten garment. Some wood-eating beetles, too, have moved indoors, particularly the furniture beetles and the death-watch beetles: here, however, the position is a little different, for man is using the natural food of these beetles—timber—and if he wants the sole use of it he must take the necessary steps to deal with any competitors.

At first sight it may appear a little hard to group the swallow, so long famed as a migrant and herald of summer, with the house-sparrow, rat and cockroach; but all of them have this in common, that they have chosen to spend part or all of their time in close association with human beings, and some of them are now so committed to this way of life that without man they would die.

4. ANIMALS, MAN AND DISEASE

MOST kinds of relationships between man and animals dealt with so far are comparatively straightforward, but in the province of health we find a triangular organization with animals acting as the carriers of diseases. In many cases the whole set-up is most intricate and one is tempted to wonder how on earth it developed. It is only natural that human diseases should have been investigated much more intensively than those affecting any other animals, but man is not alone among the higher animals in having diseases which are passed on to him by lower forms. It seems likely that in a few cases the animals concerned have long been blamed vaguely for transmitting the diseases, but it is only within the past half-century or so that modern laboratory methods have been able to work out the full stories.

In a few cases the transmission is merely mechanical; that is, the infection is carried from a sick person to a healthy one without really entering into or affecting the carrier—as a fly might pick up germs and carry them on its feet. But in most cases the process is much more complicated and only one particular kind of animal can pass on one particular kind of organism: this generally has to stay in the temporary host for some time and pass through a definite stage in its life-history. It is not just any mosquito, for instance, that transmits malaria, but one particular kind: other kinds carry yellow fever and filariasis.

Disease is normally the result of microscopic organisms

multiplying in the blood or in certain parts of the body, but a few diseases are caused by rather larger organisms: in scabies, for instance, a tiny mite—a relation of the more familiar tick—invades the skin and is solely responsible for the trouble. These animals are, as it were, the disease themselves and not the carriers of the organisms causing it, and they do not qualify for detailed discussion here. At this stage it might perhaps be well to suggest that the more squeamish readers skip this chapter.

Tapeworms are a very specialized division of a group of lower animals called the flatworms, and they are notable for their generally complicated life-histories: the final stage is normally in a vertebrate animal, which may be anything from a cold-blooded fish to a warm-blooded mammal, while the other stage or stages may be almost anywhere. Three different kinds have their adult stage in man and their other stage in animals that man eats. One of the commonest of these is the pig tapeworm: the pig picks up an egg which develops in its gut and then passes into a cyst or resting stage in the flesh: if this is eaten raw or insufficiently cooked, the young worm can continue its existence and turn into an adult tapeworm, perhaps of considerable length, in the human intestine. By rigid inspection of the pork before sale the pork tapeworm has become much less common, and there is no reason why it should not be completely eradicated. The tapeworm sometimes makes its host very thin, by taking a large part of his nourishment. The story is told of a Nigerian boy who was one of the best athletes at his school, winning most of the major events comfortably; on one of his routine inspections the Medical Officer found that the boy was harbouring a large tapeworm and dealt with it accordingly. The boy at once put on a stone in weight and failed to win another race! The moral of the story, however, is not very clear.

The beef tapeworm has a similar history: it lives its early life in the flesh of cattle, but the resting stage is nearly $\frac{1}{2}$-inch long and easily seen, so that infected beef can be rejected by inspection. The fish tapeworm has an even more complicated life, for the eggs are eaten by a tiny crustacean—one of the water fleas—in which the first stage develops: this in turn is eaten by the fish where the resting stage is finally formed in the flesh: last of all, man eats the fish, and if it has been only smoked or insufficiently cooked, he may find he has acquired the biggest tapeworm of all, for it has been known to reach a length of sixty feet.

In yet another kind of tapeworm the process is reversed: the larval stage may be in man, as well as in a variety of grass-eating animals, and the minute adult stage, under $\frac{1}{4}$-inch long, is in dogs. The cysts in man may be quite large, containing many larvae, and in certain organs they may prove fatal. The pig is also responsible for carrying another unpleasant little worm, called trichina, and passing it on to man through inadequately cooked pork: here again thorough meat inspection is a complete answer.

In many parts of the tropics a tiny water flea is the intermediate host for a round worm with an entirely different life-history. The adult grows to many inches in length and lives under the skin of humans, where it forms a small ulcer out of which it can discharge eggs whenever it comes into contact with water. These eggs enter a water flea and develop into the next stage: the host is then swallowed by man as he drinks, and so the cycle is completed by the tiny guinea worm making its way from the intestine to a suitable point near the skin. It is a thoroughly unpleasant parasite and causes much incapacity, especially as it is difficult to remove the adult worm except by the time-honoured method of winding it out, a few turns a day, on a matchstick. But it is very easy to prevent—by keeping infectious persons away from water and by boiling or filtering all drinking-water.

The life-history of many tiny worms in another group commonly called flukes or schistosomes is strangely connected with freshwater snails. The best known of the group is the liver fluke of sheep, which can cause serious losses in marshland areas: the eggs are passed with the sheep's droppings and the tiny larvae that emerge burrow into certain small snails—if they are fortunate enough to find them in time: when the end of the snail stage is reached, the young worm emerges from its host and makes its way to a blade of grass near the water's edge, which may in time be eaten by a sheep. In many parts of Africa, as well as South America and Asia, a fluke that has a somewhat similar life-history causes a debilitating human disease known as bilharzia; here, however, the tiny worm that leaves the snail is more energetic and is able to burrow through the skin of anyone bathing or working in water. The flukes live in various blood-vessels and not in the liver.

In China there is a human liver fluke which has to pass through a fish as well as a snail: in this case the fish eats the snail and then man eats the fish. With this fluke, cooking the fish would solve the whole problem, but the Chinese prefer it raw and often have no fuel to cook it with anyhow. They must also use night soil to maintain soil fertility, and as this is now applied it ensures re-infection of the snail and therefore of the fish: but it would be quite possible to deal with this manure in such a way that all eggs are killed before the compost goes on the land.

We next come to some minute worms which are actually introduced by one host into the next as the former feeds on the latter. The best known of these is the filaria worm, which causes the unsightly disease of elephantiasis in various parts of the tropics. The terrible swellings are the result of the adult worms blocking lymph ducts: at the same time these adults produce myriads of larvae, called microfilariae, so small that 600 can swim in a drop of blood. When these tiny larvae are taken up by a mosquito of a certain type (actually about a dozen closely related forms) they undergo a change taking up to three weeks and are then ready to be passed on. The pioneer research work on this disease was done by Manson over seventy years ago and he noted the extraordinary fact that during the day these microfilariae go out of the blood stream into deep tissues, reappearing at night when the Culex mosquito was about. Control should theoretically be simple—one just avoids being bitten by the mosquito! This form of filaria is native to the warm parts of most of the Old World and it has been present for many years in an area in South Carolina, presumably introduced in African slaves.

In parts of West Africa, particularly the Nigerian coast, another filaria is transmitted by a much bigger insect, called Chrysops, which is closely related to the large deer- and horse-flies that can be such a pest in some temperate zones: the story of this filaria is rather similar to the other but it normally causes only local lumps, generally known as Calabar swellings from the area where it is most common. It is in other parts of West Africa that a much smaller kind of fly—actually it is one of the midges and is usually called a buffalo gnat—transmits a tiny nematode worm that forms bumps on the skin and, what is far more serious, causes partial or complete blindness. The larval forms of the worm circulate in the blood and some of them reach the back of the eye, where they die and gradually block up the retina. In the case of all these blood-sucking flies it is theoretically possible to avoid the diseases they carry by avoiding being bitten, though in some places that is nothing but a counsel of perfection. Sleeping under mosquito nets is effective against most mosquitoes, and at the same time the breeding places of the carriers can be attacked.

All the ailments so far considered are caused by organisms that are fairly complicated, even though they are placed quite low in the general scheme of animal classification. We now come to a group of diseases caused by the most primitive of all forms of life, the tiny one-celled organisms, many of which go right inside the blood cells and can be seen only under the most powerful microscope. Malaria is perhaps the most notorious of all: for instance, it is reckoned that at least half the world's deaths every year can be blamed on malaria, either directly or indirectly, and that in India at least a million people are killed by it annually. Some authorities even consider that the collapse of the Greek and Roman Empires was caused largely by malaria: this disease has certainly been a menace in parts of the Mediterranean since ancient times and it was, in fact, in the marshes south of Rome that two English doctors finally confirmed, in 1900, that only Anopheline mosquitoes transmit the malarial parasite. The very complex life-story of this microscopic parasite need not be described here: it is enough to say that when this particular mosquito takes blood from a person suffering from malaria it infects itself and, after about a fortnight, is able to infect anybody it bites. There are actually several different kinds of malarial parasite, recognized largely by the kind of fever they cause. Several lines of attack are effective in controlling malaria. All suitable breeding places for the mosquito are oiled to kill the larvae; all dark places used for resting by day are sprayed with D.D.T.: people sleep under suitable nets to avoid being bitten: and there are various drugs for prevention and cure. Malaria probably still causes more ill-health and inefficiency than any other disease, but it is at least being conquered bit by bit: only recently, for instance, it was announced that the island of Cyprus had been completely cleared of this scourge.

Many other forms of insects and other small creatures can transmit various diseases. Ticks, for instance, carry the Rocky Mountain spotted fever in North America and the dreaded relapsing fever in Central Africa. Typhus, a disease that caused as many deaths in some armies in the First World War as did the actual fighting, is carried by that unpleasant parasite the body louse. D.D.T. almost eradicated the louse—and typhus fever—in the Second World War. The sand-fly, a tiny insect that finds most mosquito nets no barrier, passes on the sand-fly fever of the Middle East.

Sleeping sickness is the scourge of Africa and is responsible for the desertion of great areas of land: it is caused by a one-celled blood parasite, called a trypanosome, that is quite big as such parasites go and is carried only by a tsetse fly. Another tsetse fly transmits another trypanosome that causes the deadly cattle disease of nagana. Sleeping sickness develops slowly: first of all it causes general ill-health and lethargy, but death usually follows unless the patient is treated fairly early. Vast sums have been spent on this problem, but the complete answer to it is not yet known: this fly is a particularly astute one with very few chinks in its armour, so that it is very hard to attack directly. Fortunately, it seems very sensitive to changes in the vegetation cover and it has been driven from some areas by selective clearing, but it is not yet beaten and it still remains one of the biggest obstacles to human settlement in great blocks of Central Africa.

Fleas are responsible for spreading a disease which caused colossal death figures in past centuries and which is fairly called "plague": but this is, in fact, a disease of house rats, carried from one to another by rat fleas which can also migrate to human beings and give it to them if their proper hosts are killed. Plague has long been a scourge of Asia, Europe and Egypt, and it is possible that the connection between plague and rats was known to the ancient Egyptians and Hebrews. The Chinese certainly realized that when they found large numbers of dead rodents an outbreak of plague was likely to follow. Rats came from Asia, bringing with them both fleas and plague, and from about the 14th to the 17th century it periodically swept Europe. The Black Death was plague: so was the Plague of London, which killed more than every seventh person in London's population of nearly half a million. The plague bacillus was discovered in 1894, but it was only at the beginning of the present century that the true relation between fleas, rats and the human disease was worked out. It may be virtually impossible to exterminate rats, but they can certainly be kept at numbers far below the point at which plague is likely to develop.

It is a striped mosquito, often known as a yellow jack, that carries the dreaded yellow fever, the disease that delayed the building of the Panama Canal by taking a frightful toll of the workers: the cause of yellow fever was finally found to be a virus and an effective immunization was worked out during the 1930s.

House-flies receive some mention in another chapter as undesirable hangers-on. The chief charge against them is that they are carriers of the so-called filth diseases of dysentery, typhoid and cholera: these diseases can also be carried in other ways, but flies are important agents in spreading them as well as a variety of skin diseases prevalent in the tropics. In all cases the flies are mechanical carriers only, conveying the germs on their feet or mouth parts. It is little wonder that many European residents in the tropics acquire a permanent and very healthy antipathy, almost an allergy, to these dirty pests.

A small group of diseases normally affect certain animals, or groups of animals, but can also be transmitted to human beings, sometimes with fatal results. Of these the best known is probably rabies, or hydrophobia; this disease is caused by a virus and it is normally found in dogs and their very close relations, though these are able to pass it on to other animals, including humans, by biting them and injecting saliva into the wounds. The famous French scientist Pasteur did much valuable work in preparing a vaccine that could ward off the disease, even though it was administered some time after the victim had been bitten. Rabies has an extraordinarily long incubation period, so there is plenty of time to give the necessary long course of injections and thus completely stop the infection from developing. Not all "mad" dogs are suffering from rabies, especially in England, where the quarantine regulations are so effective, but it is as well to be on the safe side and give the precautionary series of inoculations after any bite by a dog suspected of being mad.

Psittacosis is another virus disease that has been in the news in the past twenty or thirty years, and at times it has received rather undeserved notoriety. At first it was thought to be a disease confined to parrots, since it was passed on occasionally to people in close contact with these birds, and it produced a condition something like what is now known as virus pneumonia; there certainly were a few human cases in which the infection had come from parrots and some of these were fatal. Now, however, it is known that parrots are by no means the only birds to carry this virus and a recent survey has shown that a high percentage of Dublin pigeons are infected, though without any ill-effects to them so far as one can see.

Among the diseases caused by bacteria there are two that primarily affect cattle but can also be transmitted to man. Malta fever is caused by a microbe generally called brucella and this illness goes by a host of names, most of which, like undulant fever, refer to its intermittent nature. It is primarily a goat disease and it is passed on to man through the milk. As it is more or less confined to parts of the Mediterranean shores and islands and is fairly easily controlled, it is not of tremendous importance, but this cannot be said of the second of these cattle diseases—anthrax. This is found in many parts of the world and it affects many kinds of animals, generally with fatal results; the organism concerned is a bacillus which causes, in effect, an overwhelming blood poisoning, but the most dangerous quality of this bacillus is that it can go into a resting stage, called a spore, which can resist considerable heat and cold and many disinfectants, and live for years and years If an animal dies of anthrax the only safe way of dealing with the carcass is therefore to burn it or to boil it in acid. There is obviously some danger to human beings wherever a sick animal is being handled or treated, and the resistance of the spores makes it liable to crop up in various industries such as the leather trade, but it is good to be able to close this rather unpleasant account with news that some of the antibiotic and sulpha drugs are completely effective against it.

Finally we can turn to a realm in which animals have made a wonderful contribution to human health. The smaller laboratory animals, by undergoing all manner of tests, make it possible to check and standardize a great range of drugs and other preparations now in everyday use against disease. The function of the larger animals is to manufacture various substances that cannot yet be prepared in any other way. Of these perhaps the best known is diphtheria antitoxin, normally prepared in horses: increasing doses of prepared germs are injected into the horse, whose blood promptly starts to make an antidote to the poison produced. After the correct interval some of the horse's blood is drawn—a painless operation that hurts the horse no more than giving blood hurts a human blood donor—and from this the valuable serum is prepared. Protection against smallpox is obtained in a rather different way: infective material is taken from a calf suffering from cowpox—a mild version of smallpox—and after preparation and standardization this lymph is used for vaccination. Our bodies are thus compelled to learn how to cope with cow-pox, and should they later come in contact with smallpox infection they can deal effectively with that. Unfortunately they do not learn this lesson permanently so that vaccination must be repeated about every five years.

5. WILD ANIMALS IN CAPTIVITY

In other chapters we have considered the domestication of animals, a phenomenon going back to the Stone Age. It is quite likely that the early peoples also kept sundry wild animals as pets, just as pets are widely kept in undeveloped parts of the world today, but with the development of civilization around the eastern Mediterranean and in the Middle East, something quite new sprang up in the keeping of wild animals in considerable numbers. The beginnings were in the very early Dynasties in Egypt and it is suggested by Loisel, who made a very detailed study of this subject, that the origin of the large-scale menageries is to be found in animal-worship. The bull symbolized the sun and later became the well-known Osiris, whilst the cow was the goddess of the moon and became Isis, the eternal night and the chaos out of which everything is born. At the same time the serpent was considered to be the spirit fertilizing Isis: its body could form the circle, always held to be the most perfect of all figures, having neither beginning nor end.

Lists have been compiled from the ancient records of the many animals which were held in reverence in those early days: they include the hippopotamus, goat, sheep, lion, cat, wolf, dog, mongoose, shrew, vulture, eagle, owl, stork and ibis, as well as various other mammals and birds that cannot be identified with any accuracy. Crocodiles and several kinds of fish, especially the Nile perch, were considered sacred in many stretches of the Nile. At times these animals were held in such reverence and so strictly protected that even the accidental killing of one of them might be punished by death. The next step was for the priests to enclose, and perhaps partly domesticate, animals which they considered to be animal gods, and thousands were kept in captivity. Many, in fact, were kept in such close confinement that the lack of sun and exercise caused serious deformation; careful examination of their mummified remains has confirmed that some of them even suffered from rickets. When some of these animals died, special periods of mourning were prescribed and they were embalmed and buried in their own tombs.

Animal training in ancient Egypt seems to have reached a level quite unknown in recent times: up to the Twelfth Dynasty, some 2750 B.C., hyenas, hunting dogs (animals entirely different from the domestic dog) and cats were all trained for hunting. It seems that the cat was already becoming domesticated and it is accepted as being a domestic animal in the Fifteenth Dynasty, some 250 years later. The cheetah was commonly used for hunting, as it still is in India, and the lion was also trained, though probably only for war.

In the days of the Ancient Empire tremendous herds of antelopes and gazelles were kept enclosed: they included the Defassa waterbuck, oryx beisa, Senegal hartebeest and several species of gazelle. This custom of animal-keeping began to become less popular after the Twelfth Dynasty and by the Eighteenth Dynasty (about 1580 B.C.) it had been almost completely lost.

Subsequently there was a revival of interest in menageries and about 1500 B.C. Queen Hatason made a Zoo at Thebes, the capital at that time, which she called the Garden of Ammon. This Zoo consisted of a great variety of north-east African animals previously unknown in Egypt, among them a giraffe. Queen Hatason's husband later imported some Indian elephants and it was presumably from these that the Egyptians got the idea of catching and training the Saharan elephants, now long extinct.

In the Middle East the Assyrian kings were both hunters and collectors of live animals. According to the most realistic carvings discovered by Layard a century ago, many of the lions hunted had first been caught and caged, and then released for the purpose, and it is obvious that wild animals must have been very numerous and the people expert in taking them alive. The tablets and other inscriptions discovered and translated in recent years reveal many facts of interest to the zoologist, even though it is not yet possible to translate with certainty the names of some animals included. In *Annals of the Kings of Assyria*, Budge and King have given literal translations of many of these archives. It seems that Assur-Nasir-Pal II reigned from about 883 to 859 B.C., in what is known as the Neo-Assyrian period, and engaged in many animal activities. He claims to have caught a dolphin in the sea (this is perhaps more likely to have been a dugong), and his people caught, among others, "herds of wild oxen, elephants, lions, wild asses, gazelles, stags, panthers"—in fact, as he goes on to say, "all the beasts of the plain and of the wilderness". He had a Zoo open to the public: more than that, he seems to have been so proud of his efforts in this line that he made his people come and see his animals. The record reads, "Their herds in great numbers I caused to bring forth and the peoples of all my land I caused to behold them." Assur-Nasir-Pal II also seems to have been competent in rearing and breeding stock, for the account goes on: "Fifteen lion cubs I carried away and in the city of Calah and in the palace of my land in cages I set them, and their cubs in abundance I caused to bring forth." Even allowing for the fact that the Assyrian Kings were given to a certain amount of boasting, they were obviously competent and keen animal-keepers.

It was some three centuries later that Daniel had his adventure in the lions' den, shortly after Babylon had fallen to the Persians. The lion, incidentally, was well known to the Hebrew peoples and there are over fifty references to it in the Old Testament, but there is no reason to believe that the Hebrews kept much other than domestic stock, even though Solomon received a shipment of apes and peacocks—and much merchandise besides—from Tarshish every three years.

Meanwhile the Greeks began to get the craze for animal-keeping: monkeys, for instance, were kept there as pets from at least the 7th century B.C., dancing bears and trained lions were known in the 4th century and at about the same time parrots were brought back from the East after Alexander's invasion of India. But the Greeks were fairly modest at home and kept their beasts—and birds—in comparatively small numbers; the Ptolemies, however, who were the Greek rulers of Egypt in the 3rd century B.C., made the most fantastic collections and the Zoo assembled at Alexandria by Ptolemy II almost defeats the imagination. The following catalogue is taken from the detailed description of the procession given by Jennison in his *Animals of Ancient Rome*. This procession, in honour of Dionysus, took all day to pass through the stadium at Alexandria, but this included sundry items other than animals. There were 24 chariots each drawn by 4 elephants, probably of the Indian species; 60 pairs of billy goats; 12 pairs of oryx; 7 pairs of saiga antelopes, each pair drawing a chariot; 15 pairs of hartebeest and 8 pairs of ostriches in harness; 7 pairs of wild asses in harness, 4 pairs drawing chariots and 4 chariots drawn by teams of 4; 6 pairs of camels laden with spices; 2,400 hounds of many kinds led by slaves. Numerous attendants carried cages containing all manner of ornamental birds. Next a whole series of domestic animals, mostly sheep and oxen of various breeds, also a large white bear, which Jennison considers to have been an albino Syrian bear rather than a polar bear. Then a wonderful array of big carnivores, including 14 leopards, 16 cheetahs, 4 lynxes and some cubs, and 24 large lions. To round off the procession were a giraffe and an African rhinoceros. These details are recorded by contemporary writers and must, presumably, be accepted as giving some indication of the numbers involved, though one figure—for a python of 45 feet long—is very much open to doubt; but there might perhaps have been giants in those days! Any modern circus or menagerie would seem quite futile by comparison.

Soon after this a different sort of spectacle developed in Rome, and in the 1st century B.C. animals began to appear in the Roman games. According to Pliny, the first elephant fights were in a show provided by C. Claudius Pulcher in 99 B.C., and he records a lion fight about the same time. The blood lust of the ancient Romans is well illustrated in an incident quoted by Seneca, when 100 lions were killed by javelin men; but Seneca records that the people had the decency to be disgusted by the wanton killing of about 20

ASSOCIATIONS BETWEEN ANIMALS AND MAN

elephants, some of which fought very gamely. Julius Caesar also gave a number of extravagant shows, one of them including 400 lions. It seems that many of the lions came from south-west Asia, where they were still very numerous. The bloodshed in the early days of the Roman Empire was nauseating; it is recorded, for instance, that 9,000 animals were slaughtered in 100 days when the Colosseum was opened. It was at this period that the Christians were thrown to the lions and other wild beasts, and the whole thing was planned to be dramatic and drawn out. At the same time Plutarch records that many animals were being trained and, as in Egypt a few centuries before, the variety and numbers of such trained animals were stupendous.

Nero, in the middle of the 1st century A.D., kept up the tradition, and one of his spectacles is said to have included a fight between cavalry and 400 bears and 300 lions; bull-fighting was popular and he also organized races for camel chariots. But Nero also seems to have been genuinely interested in animals and kept a great variety of them around his house. The games went on until early in the 5th century, when the Roman Empire was divided, but the beasts killed seldom again approached the colossal numbers provided by some of the earlier Emperors. This was due, at least in part, to the increasing difficulty in obtaining supplies of suitable animals through the governors of the various Roman Provinces. However, the beasts were not always provided only to be slaughtered, for some Romans, both high and low, were obviously very fond of their animals. The Emperor Caracella, for instance, was particularly interested in lions and kept a number of them which he took on his journeys; one of them, Acinaces or Scimitar by name, even ate with him and slept in his bedroom.

It is perhaps worth while to enumerate some of the menageries of other Roman Emperors, details of which are available in the official records. Octavius Augustus, 29 B.C. to A.D. 14, had 3,500 head, including 420 tigers, 260 lions, 600 assorted African carnivores, 1 rhinoceros, 1 hippopotamus (the first to be seen in Rome), seals, bears, elephants, eagles, 36 crocodiles and a colossal snake. Trajan is said to have had 11,000 wild and tame animals (98–117). Heliogabole (218–222) went in for variety, and as well as the usual lines he had many hippos and ostriches, rats and mice in their thousands, and even scorpions. Probus (276-282) had 1,000 each of ostriches, deer and wild boars as well as chamois (a very difficult animal to keep), giraffe and wild sheep.

Hortensius had an open-air menagerie in his park at Laurentium; one of his servants, dressed like Orpheus, blew a hunting horn to call his large collection of animals together as his guests assembled to dine. Many other scraps of information in the classical records show that the animal trainers of those days were fully equal to those working today. A wall painting unearthed in the buried city of Pompeii shows a keeper leading a giraffe around by a halter; few people in this country have seen such a sight, though one giraffe—the famous Twiga, now at Whipsnade—was actually broken to the saddle and ridden fairly regularly in Kenya as a colt. Again, it is now considered quite exceptional to breed elephants in captivity (largely because of the difficulty and danger of keeping adult bulls), but it seems that the gestation period of the elephant was known to Aelian and it is fairly certain that some performing elephants famous in the latter part of the reign of Augustus had been born in Italy. Another wall picture at Pompeii depicts an elephant mother and young which are obviously tame animals; more than that, the general shape and particularly their ears show them to be of the African species. In this connection it is perhaps worth noting that it was at the Rome Zoo that elephants were first bred in captivity after the Second World War; more than that, the mother refused to care for the calves and two, born in 1948 and 1950, have been hand-reared successfully.

Other points in the old records seem to have a modern ring about them: the lawyers, for instance, defined the precise legal position in the case of damage done by domesticated and wild animals, and even in Roman times lions and leopards were dutiable, though other groups had exemption, and it apparently paid to get animals collected in the overseas Provinces. But with the collapse of the Roman Empire all this came to an end, and it was not until the Middle Ages that anything like a menagerie was again seen in Europe.

Compared with the ancient civilizations of the Middle East, very little is known about animals in old China. Chinese hieroglyphics show leopards, rhinoceroses and elephants, as well as the usual domesticated animals, and it may well be that it was the custom to keep many kinds in captivity, but the first definite mention of captive animals is an inscription on a vase dated about 1600 B.C. instructing a minister to go and collect four pairs of different animals for the Emperor Keng-ou. Confucius refers to a deer house constructed in marble by the Empress Tanki about 1200 B.C., and about 200 years later the Emperor Wen Wang made an animal park of nearly 1,000 acres between Peking and Nanking. It may have been about this time that one of the Chinese Emperors first enclosed the herd of deer now known as Père David's, whose extraordinary history is given in an earlier chapter.

After an interval of over 2,000 years, with little or nothing of interest being recorded about China, Marco Polo gives us a description of his visit in 1271. He found Kublai Khan keeping many wild animals, including lions and tigers walking about in complete freedom. In the gardens he had several varieties of deer, including a white form, and also a large artificial fishpond. It is difficult to suggest any dates but the Chinese have long been specialists in fish culture, both ornamental and utilitarian; in the former class come the goldfish, with their fantastic variety of forms, while in the latter must be included the highly successful mixed communities of food fish which, with the soya bean, have made human survival possible in China. In these skilfully balanced fishponds some three or four species occupy quite separate niches and together form a most efficient food-producing unit where nothing is wasted.

In India also it was usual to keep captive animals, especially tigers and elephants, but there is nothing to suggest that foreign kinds were kept. The Mogul Emperors and the Kings of Siam kept very large numbers of elephants, primarily for ceremonial purposes, though it was also the custom to stage elephant fights. Cheetahs were trained for hunting in early times and are still so used today. It seems likely that in the past ten centuries there has been less change in animal-keeping in India than in any other part of the world.

On the other side of the world we find some tradition of animal-keeping in the civilizations of the Incas of Peru and the Aztecs of Mexico. There are no written records and we can only speak with certainty of the position when the Spanish conquerors arrived. The Incas worshipped a number of animals, including the puma, llama, condor and eagle, and had the llama, alpaca and guinea-pig in domestication, but they did not keep many wild species in captivity. The Aztecs under the Emperor Montezuma, on the other hand, had extensive collections when Cortez arrived in Mexico in 1519. There were large flight aviaries, complete with pools, and no less than three hundred men and women were detailed to look after the birds in them; their tasks included catching large numbers of insects for those birds that needed them, so that the aviaries probably housed some of the insectivorous kinds that are among the most difficult to keep in captivity. Unfortunately there does not seem to be any account of the type of construction of these large flight aviaries. The Aztecs also kept reptiles of all kinds, carnivores and birds of prey; llamas and vicuñas are also mentioned, which can only have come from South America. These collections were all housed in gardens made beautiful with a profusion of flowering trees and shrubs.

Turning back to Europe we have to record a gap in the tradition of animal-keeping between the end of the Roman Empire and the Middle Ages, a gap of six or seven centuries. The first menagerie of any size on the Continent of which there is any record was founded in the 13th century by Frederick II, King of the Two Sicilies, and he got together a considerable collection which he used to take around with him. About that time some of the Governors of the Italian Provinces started keeping wild animals in captivity and in Rome a den of lions was established at the front of the Capitol; it was these lions to which a monk was thrown in 1328 by Louis IV of Bavaria when he captured the city. The last of them was killed in 1414 after escaping and killing a child, and the dens were then used to house only goats, pigs and geese.

The famous bear-pit at Berne in Switzerland is generally said to have been started in the 12th century by Berthold V, Duke of Zachringen, but it may perhaps go back many centuries farther than that, for very ancient bronze figures have been discovered in the neighbourhood which show a bear going to take food from the hand of a goddess. The first authentic mention of the pit was in 1480, when it contained a solitary bear. In 1513 a company of Swiss mercenaries returned to Berne bringing with them as a trophy a brown bear which they had found among the enemy spoil. The bear was kept as a souvenir of the victory and others were caught in the mountains as companions for it. Ever since then the pits have had their bears and they are one of the sights that all visitors see; the number of photographs, paintings, carvings and other portrayals of bears sold annually in Berne must be quite beyond computation. Wild bears have not been seen in Switzerland since 1914, so that new stock must now be brought in occasionally from abroad.

The history of menageries in Great Britain is fairly easy to trace. The Romans certainly brought some domesticated animals with them, but there is nothing to suggest that they tried to establish any menageries; they are more likely to have regarded Britain as a collecting area from which interesting specimens should be sent back to Rome. The first *bona fide* menagerie of wild animals seems to have been the one established in the 11th century by William the Conqueror at his residence at Woodstock. Early in the following century Henry I had a fairly extensive collection, just the sort of thing that a good travelling menagerie would have today, including lions, leopards and lynxes, and various birds; there were also camels, presumably of the Arabian kind. It was about this time that the Norman Barons made the parks of Cadzow, Chartley and Chillingham, in which they enclosed the herds of cattle which survive today under these names; although there is argument about their origin it is generally agreed that the form of these cattle has not changed appreciably for some seven or eight hundred years.

Henry III maintained the animal-keeping tradition by having a white bear, presumably though not certainly a polar bear, at his court in London in 1251, but he made the people pay for it and it is recorded that the Sheriffs allowed 4*d*. per day for feeding the bear and paying the keeper. An even more interesting point is that the bear was allowed to catch some of its food in the Thames—an indication that it was a very different river in those days. The city had to supply a muzzle and an iron chain as well as a long cord on which it went fishing. A few years later Henry received an elephant as a present from his brother-in-law, Louis IX of France, the first to be seen in England. The Royal menagerie had been moved to the Tower during the 13th century, but it remained a charge on the people of London until the following century. For a long time numbers remained fairly small and little is recorded about the Tower collection until the 17th century, when James I increased its size, though he seems to have used the animals principally for staged fights.

About one hundred years later Queen Anne improved the conditions under which the collection was kept and it then consisted of 11 lions and 2 leopards or tigers, 3 eagles, 3 eagle owls and various other animals. George II added bears and monkeys, and it is recorded for the first time that the exhibition was open to the public on the payment of 3*d*. per head. In 1802 there were 4 lions, two of which had been bred in the Tower, a black leopard from Malabar, a Bengal tiger and a bear. Twenty years later George IV ordered quite a lot of work to be done to the cages, but in 1840, the year after his death, all of the animals were got rid of, some going to the recently formed London Zoo and some to Dublin.

Meanwhile the Zoological Society of London had been founded in 1826 and incorporated by Royal Charter in 1829; the Gardens, occupying a tiny fraction of the present site in Regent's Park, were first opened to the public on 27th April, 1828. This was not the first of the big European zoos to be founded, for the Jardin des Plantes menagerie in Paris dates from 1732, and the botanical section over a century earlier still. Many famous zoologists have been connected with the Paris Zoo, including Buffon, Daubenton, Cuvier, Geoffroy and Milne-Edwards. Many animals were named by or in honour of these scientists; the extraordinary and quite unique aye-aye from Madagascar, for instance, bears the generic name of *Daubentonia*, while one of the British bats is generally known as Daubenton's. The Schönbrunn Zoo in Vienna is the next oldest, having been started in 1752 under Maria Theresa. The first half of the 19th century was a time of great zoological activity in western Europe, and most countries acquired a Zoo run either by a Society, as in London, or as a municipal undertaking, as in Paris, where the menagerie is an integral part of the national Museum. Dublin got its

Zoo in 1830, Amsterdam in 1837, Antwerp in 1843, Berlin in 1844, Copenhagen in 1859 and Basel in 1874.

All of these gardens had more or less the same objects, and these are roughly described in the following paragraph from a circular prepared by Sir Stamford Raffles, the true founder of the London Zoo, in 1825:

> A collection of living animals belonging to the Society will be established in the vicinity of the metropolis; to which the Members of the Society will have access as a matter of right, and the public on such conditions as may be hereafter arranged.

All of these Zoological Gardens exist today, one or two of them, alas, still more or less in the state which they reached before the end of last century, but the bodies concerned with the administration of most of them realize the need for keeping up to date and making use of the latest discoveries bearing on all aspects of animal management and exhibition. Those Zoos enumerated—and a number of others also—are described generally as non-profit-making cultural undertakings; that is, they are run for the general good and not for anyone's personal profit. Should a surplus be made on any year's working this money is put back into the undertaking, whether as a reserve, or used for rebuilding, or in some other way. This century has also seen the foundation of many smaller Zoos with the partial or sole object of making money for the promoters. Some of these can be regarded as the direct descendants of the older travelling menageries, but some have started off as purely private collections and have subsequently been made accessible to the public; the admission fees then help in the maintenance of the collection. A more important development is the establishment, by the Zoological Society of London, of Whipsnade Park, where a smaller range of animals is shown in conditions much more approximating to freedom: this park, about thirty-three miles north of London, was opened in 1931. Somewhat similar principles are applied in the famous Munich Zoo under Dr. H. Heck and in some of the numerous Zoos that have been founded in the United States of America.

There are some folk who object on principle to the keeping

of wild animals in captivity, irrespective of the conditions in which they are kept. These objections are largely based on misunderstandings and on trying to put oneself in the animal's place, arguing rather on these lines: "That animal is in a cage: I should hate to be in a cage: therefore that animal hates being in a cage and is unhappy." It is true that some animals —the minority—refuse to settle down in captivity, but many of them, including all those that have become firm Zoo favourites, not only settle down but appear to enjoy life. How do we know that? We cannot get right inside the mind of an animal but it seems fair to assume that if the animal feeds well and keeps in excellent condition, if it seems genuinely glad to see human visitors and, even more, if it breeds freely, then it is happy and contented. In many cases the animal, be it bird or mammal, comes to know for the first time what it is to have freedom from fear, want and competition, and this may be ample compensation for the partial loss of freedom of movement. It is interesting to note that most birds and mammals have a distinctly longer expectation of life in captivity and may even die of old age: that is a condition virtually unknown in the wild. It is also worth noting that few people object to the keeping in captivity of cold-blooded animals— fish, reptiles, insects and so on: this is probably because they appear so far removed from the human form that it is hard to credit them with the human feelings so often ascribed to birds and mammals. These objections, however, are made by the small minority, and to most people Zoos are places of wonderful entertainment and instruction: they offer the only opportunity most of them will ever have of seeing foreign animals—and even many from their own homeland. Properly used, Zoos are of great scientific value: ideally there should be the closest co-operation between Zoological Gardens on the one side and Veterinary Schools, Zoological Departments and Museums on the other, and this should be to their great mutual benefit.

6. ANIMALS IN RELIGION, FOLKLORE AND SUPERSTITION

FINALLY we come to a whole range of rather nebulous relationships between animals and man that can be described roughly as being in the realm of ideas. It is hard to know just where to begin and almost impossible to decide where to leave off. As regards religion, a detailed historical treatment would perhaps be the most logical, for this association started in the early ages of man, when polytheism began to develop from the original monotheism. (I find myself in agreement with the theory recently confirmed by some archaeologists and anthropologists that monotheism is older than polytheism and that at the beginning man worshipped the one true God.) In the beginnings of civilization in the Old World, in both Egypt and in Mesopotamia, images were carved in the likeness of animals, sometimes a mixture of two quite different animals, and these were apparently considered to be intermediate between the gods and men. The colossal bull-like statues carved by the Assyrians, some of which can now be seen in the British Museum, are thought to have been guardian spirits on duty outside the Palace.

The next stage was to look upon animals as gods, or to make images of animals and regard them as the earthly dwelling-place of the gods. Egypt had many such animal gods and their possible bearing on the domestication of animals is dealt with elsewhere. The best known of the Egyptian animal deities were the wolf, falcon, dog Anubis (this name is seen still in *Papio anubis*, the Latin name for the dog-faced baboon), ram, ibis, bull, lioness and the sun-god Khopri in the form of a scarab beetle. Some of these strange forms had bodies of one animal and heads of another. Hathar, for instance, was a cow with a human face: Month was the hawk-headed man, while Sakhini was the lion-headed goddess of war.

In the monotheistic religion of the Hebrews animals occupied a very different place, and some were the subject of elaborate sacrifices: domesticated animals alone were used—cattle, sheep, goats and pigeons. These regular sacrifices, begun at the Exodus, were the central feature of their religion. Animal-worship and the veneration of animal images were absolutely forbidden, and the Hebrews were made to regard as unclean those animals which were deified by the heathen peoples around them. Sacrifices are still part of the Jewish ritual, but they find no place in Christian worship. In many other religions, however, placatory sacrifices are still the rule, as well as other offerings connected with fertility cults.

A list of animals worshipped in the past and at the present day would be lengthy. In Siberia, for instance, it was formerly the custom to capture a bear and keep it for some time while certain rites were performed: after that it was fattened up, killed and eaten. In Ancient Greece the goddess Artemis was connected with the bear. The cow has been venerated by the Hindus since something like 1500 B.C., and Gandhi claimed that this was the distinguishing mark of the Hindu. Dog-worship is rare today (though there are perhaps examples of it in ultra-sophisticated communities!), but it was common at one time, and in ancient Egypt the dog was considered a very desirable goal for the human soul after death. Dagon, the god of the Philistines, seems to have been a fish with human head and hands. The goat was popular in Greek mythology and the gods Pan, Dionysus and various others were partly in the form of a goat.

The North American Indians had sundry sacred animals, including the cow and the hare. In the hinterland of West Africa the Moshi peoples to some extent regarded the hare as sacrosanct, and when I was working in that part I was constantly begged not to shoot a hare and thus bring trouble to my party. The hare is the animal hero in other parts of Africa also, though generally known as a rabbit, and it is probable that the Brer Rabbit of North America is nothing but this African hare as remembered by the slaves. Various types of monkey are considered sacred, especially the Hanuman langur, which is the temple monkey of India. The serpent cult is perhaps the most widespread of all, though it is best known in Dahomey, in French West Africa. Other African tribes indulge in ritual dances involving the handling of live cobras and other poisonous snakes, and in parts of North and Central America rattlesnakes are similarly used. It is possible that the snake-charming in Egypt and India, mostly using cobras, had its origin in serpent-worship, but today it is nothing but a money-making device.

The leopard cult in West Africa—particularly Nigeria and Sierra Leone—is quite another matter: these societies are, in effect, little other than murder societies, but the general idea is to get the rather ignorant populace to believe that the members literally turn into leopards by night. Even now there are reports of "leopard" murders from time to time.

Totemism is a phenomenon found among many different peoples, though it is best known among the North American Indians, and the name actually comes from one of the Algonkin dialects. It is mystical rather than religious and the totem can be described as the hereditary mark, generally consisting of a figure of an animal, after which the group is named. The Indians often had three main divisions, as for instance wolf, turkey and turtle; in the wolf division all totems were mammals, in the turkey all were birds, and in the turtle all were reptiles or amphibians. Rules varied a great deal, but it was generally compulsory to marry a different totem within the same division, thus ensuring incidentally that there is not too much in-breeding.

A rather similar organization is found among the Australian aborigines, but there are generally two major divisions such as hawk-eagle and crow, or white cockatoo and crow: within these there are the totem classes, such as wombat, kangaroo, dog, emu, owl, frog and so on. The patrilineal divisions on the Gold Coast are rather akin to totems, but the name given to each is a proper name and the animal connections are in the food taboos attached to each.

The famous White Horse of Uffington carved in the chalk of the Berkshire Downs is considered by some authorities to be of totemic significance. It has been given various meanings at different periods but it is thought to date back to the early Iron Age—that is, about the 1st or 2nd century B.C. The Horse is 365 feet long and very diagrammatic in form.

Animals also occupy a large place in superstition, or perhaps it is more correct to say that they did so in Great Britain and western Europe in the Middle Ages. It is not that we are any less superstitious, for a recent enquiry has shown that some 25% of British people are to some extent affected, but that the emphasis has changed to more material things such as ladders, salt and the "unlucky" number thirteen. In Act IV of *Macbeth*, for instance, there is some indication of the sort of ingredients that were used to make a mediaeval witch's brew. All the main British amphibians and reptiles went in: the toad, frog and newt: the grass snake (called here the fenny-snake from its liking for wet places), adder, slow-worm (blind-worm) and lizard. But several mistakes common to this day are apparent, for the toad has no venom and the blind-worm has no sting. The tabby cat and the hedgehog are mentioned as being in the background, but they were not part of the recipe; except for the goat, more exotic larger animals were needed—the dragon, wolf, shark and tiger, with baboon's blood for cooling. The only bird mentioned is the owl, and this is associated with the occult in many lands, presumably because of its silent nocturnal habits.

Comparatively few British animals are habitually connected with superstition today. The black cat stands now for good luck, but at one time the close association of cats with witches was not at all healthy for the cats. The raven is generally held to be a bringer of ill-tidings, dating back, perhaps, to the raven released by Noah, but in a later incident in the Old Testament the ravens more than retrieved their reputation by supplying food to Elijah. Its sepulchral voice may have something to do with it, as in the case of the pied crow of West Africa. These birds were a danger to young chickens, and I often used to shoot them, only to find that no tribe would eat them: this was explained not by their partially carrion-eating habits but by their voice, which was considered too human-like.

With the magpie it is a matter of numbers, and there are several versions of the rhyme which commonly starts "One for sorrow: two for joy".

In the Gold Coast it was a little difficult at times to distinguish superstition from physical fear resulting from things believed honestly but in error. The chameleon was everywhere dreaded: everyone fully believed that it was deadly poisonous, and whenever I persuaded a chameleon to bite on my finger to prove the contrary there was always a great stir. But there was more to it than just physical fear of being bitten, for the chameleon is, in all conscience, a queer enough creature: the skin colour can change, the tail is prehensile, the feet are unique, and, most disturbing of all, the eyes are quite independent and can fix firmly two widely separated objects. I don't really blame the African for being scared of the chameleon.

Folklore seems invariably to have a lot to do with animals and this is certainly the case in West Africa. With the Ashanti people, for instance, this is reflected very plainly in the proverbial sayings in which their language is so rich: this language, generally called Twi, was reduced to writing by missionaries about a century ago, and as until recently there was a very low percentage of literacy, these sayings were normally carried in the memory and were very widely known. My work as a Forest Officer involved holding palavers in the villages, for I had the often difficult job of putting over the policy of forest conservation (very much in the people's real interest), and I found a knowledge of these proverbs invaluable; they were rather like trump cards, and if the correct saying for a particular point could be produced I won that point.

The animal hero of Ashanti was a spider called Ananse, and there was a wealth of folk-stories called Anansesem, or the words of Ananse, in which he featured. The actual spider identified with Ananse was a large harmless web-spinner with the Latin name of *Nephila*. Ntikuna, said to be Ananse's son, was an odd-looking wall spider with no common English name. One proverb says that one does not tell Anansesem to the son of Ananse and it has much the same meaning as our admonition "not to teach one's grandmother to suck eggs".

Other sayings involving animals have their exact parallels

in well-known English proverbs: "The pot calls the kettle black" is more picturesquely rendered "The vulture says that the civet cat stinks"; "One does not get blood out of a stone" is expressed naturally by "A tsetse fly sits in vain on the back of a tortoise"; "He whom a snake has bitten afterwards fears a worm" is much more incisive than "Once bitten, twice shy". Perhaps the Ashantis are greater realists than we when they say "A bird in the hand is worth ten in the sky". Many of these sayings contain pungent truths which are not so well known in proverb form in English, as, for instance, "No one sells a laying hen without a very good reason" and "When the rooster gets drunk he forgets all about the hawk" (the rooster's main job is considered to be that of watchdog). Another interesting one is "The tortoise says that haste is good but so is deliberation". Perhaps the most familiar of all the Ashanti proverbs should be included in this group; it runs, "A boy can crack a snail but cannot break open a tortoise". This proverb is, in fact, so familiar that the speaker can only say the first part and the audience always finishes it as a chorus.

But many of these sayings would not be ranked as proverbs in English, even though they express interesting or amusing thoughts; there is very little inhibition in this type of speech and some of them are best left untranslated. The elephant features frequently and some of the proverbs which compare it with the tiny royal antelope are strikingly reminiscent of the old English tale of the conflict between the eagle and the wren. The Ashantis say, "The elephant may be big but the little royal antelope is the King of the Forest". It is worth noting that the royal antelope, the smallest antelope in the world, got its English name from this proverb which was told to the collector of the first skins sent to England some 200 years ago. But another proverb also says that "Although the elephant is enormous (the word used means literally 'big for nothing') it is the little royal antelope that makes the best soup". Other elephant proverbs are: "One man can kill an elephant but many eat it"; "You may make a bag of an elephant hide but how can you fill it?"; "It is not just because it eats more food that the elephant is bigger than a dwarf antelope".

The baboon is also mentioned frequently, and the following proverb refers to a feature very well known in this group of monkeys: "The baboon trots along like a proud man but his hind quarters are red nevertheless". Oddly enough the chimpanzee seems to have no mention at all in these proverbs, possibly because it is rather scarce in most of the Gold Coast. The domestic fowl, found in every compound, however humble, is rather naturally the subject of many sayings. "The fowl drinks water a little at a time" is one of the best known of all. Another says that when the fowl drinks it first shows the water to God—that is, it says "Thank you, God". Almost all the domestic animals have frequent mention, though few of the sayings are very striking; one, however, is worth quoting as expressing a belief by no means unknown even in civilized lands, "The sheep says that it is looking at the leopard so that it may bear a lamb looking like one".

At one time the Gold Coast forest peoples were expert in making cats' cradles or string figures, known there by the picturesque term "wise men's knots". It seems likely that these cats' cradles have now been almost completely forgotten, but when I first went to Ashanti in 1934 I found that the old men in the villages still knew many versions and I was able to collect and record about fifty different varieties. These were just as much a part of the folklore as were the Anansesem, and I found that many animals featured in them: such, for instance, as the butterfly, mosquito, leopard, bongo, Gaboon viper and dwarf mongoose. Several figures illustrated this particular mongoose and they all refer to a belief, current from the Gold Coast right across Africa to Uganda, that it lies down and exposes its white fleshy anal scent-glands in the hope that a fowl will mistake them for grubs and thus be caught. For some reason this mongoose story was thought to be rather indecent and there were always roars of laughter when the villagers found that I knew it.

Some animals lend themselves to use in idioms—the goose is a good example of this. As long ago as 1547 a simpleton was being referred to as a goose; presumably a domestic goose, for a wild goose is anything but a fool. But since then the goose has been incorporated in several quite unrelated semi-proverbial sayings, including: "All his geese are swans"; "To cook one's goose"; "To kill the goose that lays the golden eggs"; "To say 'Boo' to a goose".

This, presumably, is the chapter where mention could be made of the animals in Greek and other mythology, such as the Centaurs and Pegasus, which, with many of their contemporaries, are now best known by the constellations of stars named after them. Heraldry, also, could find a place here, for many of the animals appearing on arms are so formalized that they have little relation to real animals: the martlet, for instance, is now officially described frankly as "an imaginary bird without feet, born as a charge", and the British unicorn is just as mythical. Inn signs, too, might call for some mention, for animals frequently form the main device: the Dun Cow, the Red Lion, the Bald-faced Stag (a stag with a white blaze on the face), the White Hart, the Coach and Horses and the Nag's Head—all these signs, interpreted very freely in many cases, can be seen in many parts of Great Britain. But by this stage the association between animals and man has become so slender that it hardly exists and it is, therefore, time to stop.

BIBLIOGRAPHY

ALLEN, G. M. (1942). *Extinct and Vanishing Animals of the Western Hemisphere*. Washington. American Committee for Internat. Wild Life Protection. Spec. Publ. No. 11.

BERTRAM, G. C. L. (1950). "Present Status of the Marine Animals." *J. Soc. Pres. Fauna Emp.*, N.S. No. LXI.

BUDGE, E. A., WALLIS and KING, L. W. (1902). *Annals of the Kings of Assyria*. British Museum.

BUSVINE, J. R. (1951). *Insects and Hygiene*. London. Methuen.

DARLING, F. Fraser (1947). *Natural History of the Highlands and Islands*. London. Collins.

FITTER, R. S. R. (1945). *London's Natural History*. London. Collins.

GEE, E. P. (1951). "The Chillingham Wild White Cattle." *Scot. Zoo and Wild Life*, III (4).

HAHN, E. (1896). *Die Haustiere und ihre Beziehungen zur Wirtschaft des Menschen*. Leipzig. Verlag von Duncker and Humblot.

HARPER, F. (1945). *Extinct and Vanishing Animals of the Old World*. New York. American Committee. for Internat. Wild Life Protection. Spec. Publ. No. 12.

HARTING, J. E. (1880). *British Animals Extinct Within Historic Times*. London. Trübner & Co.

JONES, F. Wood (Ed.) (1951). "A Contribution to the History and Anatomy of Père David's Deer." *Proc. Zool. Soc. Lond.*, Vol. 121, p. 319.

KEARTON, R. and C. (1897). *With Nature and a Camera*. London. Cassell & Co.

LOISEL, G. (1912). *Histoire des Ménageries de l'Antiquité à nos Jours*. Three vols. Paris. Henri Laurens and Octave Doin et Fils.

LUCAS, A. *Ancient Egyptian Materials and Industries* (Chap. III, Animal Products). E. Arnold & Co. London. 3rd. Ed. 1946.

MATHESON, C. (1932). *Changes in the Fauna of Wales in Historic Times*. Cardiff.
(1942). "The Passing of a Fauna." *J. Soc. Pres. Fauna Emp.*, N.S. No. XLVI.

MENZIES, J. R. (1951). "Man and the Zebra." *Oryx*, 1 (3).

MOHR, E. (1949). "Development of the European Bison during the Last Years." *J. Soc. Pres. Fauna Emp.*, N.S. No. LIX.

THÉVENIN, R. (1947). *Origine des animaux domestiques*. Paris. Presses Universitaires de France.

ZABINSKI, J. (1949). "Twenty Years of Bison Breeding in Poland." *J. Soc. Pres. Fauna Emp*, N.S. No. LIX.

ZEUNER, F. E. (1951). "The Cat." *Oryx*, 1 (2).

INDEX

Acclimatization Society in N.Z., 65
adder, 188
Africa, Central, 30, 31
——, East, 28, 30
——, North, 26, 28
——, South, 28
——, West, 31
alpaca, 110
ambergris, 151
amphibian, 145, 162, 187
Angora goat, 90
animal fats, 146
—— worship, 173, 185
ant, 135
antelope, 142, 174
——, royal, 190
——, saiga, 176
anthrax, 171
antlers, 16, 150
ants and green-fly, 84, 153
—— in rat control, 73
Anubis, 185
apes, 142, 175
aquaria, 136
ass, domestic, 85, 106
——, wild, 28, 106, 175, 176
Aububon Society, 38
auk, great, 22
aurochs, 16, 46
Australia, 44, 63, 72, 157
avocet, 18
aye-aye, 182

Baboon, 35, 188, 190
badger as food, 142
—— -baiting, 133
—— -digging, 133
—— hair, 150
bandicoot rat, 155
banting, 88
bat, Daubenton's, 182
—— guano, 147
bats, fruit, 66, 140
——, insect-eating, 71, 161
bear, 142, 175, 176, 177, 181
—— -baiting, 132

bear, brown, 14, 180
beaver, 15, 37, 95
Bedford, Duke of, 46
bee-keeping, 101
beef, 142, 165
beeswax, 102
beetle, Colorado, 57, 66
beetles, wood-eating, 163
bengalee, 117
big-game hunting, 131
bilharzia, 167
bison, American, 36, 37, 46
——, European, 37, 46, 47
bittern, 18
blauwbok, 28
blesbok, 30
blindness, 168
blindworm, 188
boar, wild, 15
bone, 150
bones, fish, 150
bongo, 33, 191
bontebok, 30
botanical collecting, monkey used in, 123
brer rabbit, 187
budgerigar, 85, 117
buffalo, water, 88
bull-baiting, 128, 132
—— -fighting, 132
bunting, 157
butterflies, 69, 191
buzzard, 21, 62, 160

Cabbage worm, 74
cactus, 77
camel, 85, 108, 147, 170, 181
——, Arabian, 108
——, Bactrian, 108
—— chariot, 177
Cameroons, 31
canary, 116
capercailzie, 17, 21, 61
caribou, 92
carp, 119, 145
cash-crops, 27, 34
Cashmere goat, 90

INDEX

cat in Australia, 45, 64
——, domestic, 114, 146, 173, 174, 188
——, Siamese, 115
—— as mouse control, 71
—— in New Zealand, 65
——, wild, 20, 62
caterpillars, 69
catgut, 150
cat's cradles, 191
Cattle, British Park, 87, 181
Centaur, 192
centipedes, 56
chameleon, 189
chamois, 178
cheetah, 123, 127, 131, 176, 179
chimpanzee, 123
chinchilla, 95
Christianity in Africa, effects of, 27
Chrysops, 167
chub, 145
citrus pests, 74, 75
civet, 123, 190
climate, effect of, 22, 69
cobra, 162, 187
cochineal, 151
cockatoo, 187
cock-fighting, 128, 133
cockles, 138
cockroach, 72, 162
cocoa crop, 34, 84
collectors, 18
condra, 180
coney, 135
cormorant, 123, 146
corncrake, 22
cow (see ox, domestic)
—— -pox, 172
cowrie, 151
coypu, 94
crab, mitten, 58
crag-martin, 160
crane, 135, 146
cricket, house, 163
Criollo horses, 106
crocodile, 144, 146, 147, 148, 173, 177
crossbill, 17
crow, 20, 187
——, carrion, 20
——, pied, 140, 188
crustaceans, 138
cuckoo, 84
curlew, 22

cushion scale, cottony, 74

DEER, fallow, 128
——, Florida key, 40
—— -forest management, 14
——, musk, 151
——, Père David's, 45, 46, 179
——, red, 16, 17, 65, 128
——, sambar, 65
deerskin, 148
diphtheria, 172
dodo, 41
dog, domesticated, 85, 112, 124, 127
—— as meat, 140
—— in Australia, 45
—— in worship, 173, 186
——, hunting, 174
——, wild, 112
—— -fighting, 132
dolphin, 175
domestic animals and human migration, 67
domesticated animals, definitions of, 83
donkey (see ass, domestic)
dormouse, fat, 59
dove, Barbary turtle, 118
——, rock, 117
dromedary, 108
duck, domestic, 100
——, muscovy, 100
dugong, 42, 175

EAGLE, 20, 173, 180, 182
——, golden, 124
——, sea, 19
—— owl, 182
eel, 145
——, Roman, 145
egg collectors, 19, 21
eggs as food, 143
egret, American, 39
——, snowy, 38
eiderdown, 149
elephant, 110, 122, 174, 175, 176, 177, 178, 179, 181, 190
elephantiasis, 167
elk, 15
entomology, 136

FALCON, 185
falconry, 124, 127, 131

INDEX

farming, effect of, 20
——, ——, in West Africa, 32
feathers, 149
fens, 18
ferret, 66, 71, 73, 116, 130
Field Study Centres, 135
filaria, 167
finches, 157
fish as food, 145, 179
—— -farming, 145
—— in worship, 173
—— -keeping, 136
fishing, 131
——, results of, 22, 23
fishponds, 119, 136, 179
Florida, 40
fluke, liver, 166
food taboos, 142
Forest Reserves, 34
forestry, biological control in, 76
fowl, domestic, 96, 142, 190
fox, 20, 64
——, Arctic, 95
——, silver, 95
—— -hunting, 129, 134
frog, 188
——, clawed, 122
——, edible, 62
——, marsh, 61
frogs as food, 145
fulmar, 143

GALAPAGOS Islands, 44
gamebirds, 130
gamekeeper, 18, 19, 20
gamekeeping, results of, 20
gannet, 25, 143
gasworks, effects of, 24
gaur, 88
gayal, 88
gazelle, 146, 148, 174
gecko, 72, 135, 162
——, Brooke's, 56
giraffe, 174, 176, 178
gnat, buffalo, 168
gnu, white-tailed, 30
goat, 54, 85, 86, 90, 146, 171, 173, 176, 180, 186, 188
—— introduced to Guadaloupe, 63
—— —— New Zealand, 65
goatskin, 148
godwit, black-tailed, 18

Gold Coast, 31
goldfish, 120
goose, 99, 180, 191
——, Chinese, 99
——, Egyptian, 99
goshawk, 124, 130
green-fly, 84, 151
greyhound, 127, 128, 134
grouse, 130, 131
Guadaloupe, 63
guano, 146
guinea-fowl, 98, 131
—— -pig, 121, 141, 180
—— -worm, 166
gulls, 159
gut, 152

HALIBUT, 23, 145
hamster, golden, 122
Hannibal's elephants, 111
hare, 129, 186
harrier, marsh, 18
——, Montague's, 18
Hart, White, 192
hartebeest, 174, 176
hawk, 20, 190
—— -eagle, 187
hedge-sparrow, 158
hedgehog, 188
heraldry, 192
heron, 124
hippopotamus, 146, 151, 173, 177
hogskin, 148
honey, 102
horn, 151
horse, 85, 103, 172
—— in sport, 128, 134
horsehair, 150
horses, draught, 105
hounds, 176
house-fly, 162, 170
—— -sparrow, 156
huanaco, 110
humble bee, 72
humming-bird, 149
hydro-electric schemes, 24
hydrophobia, 170
hyena, 140, 174

IBIS, 186
ichneumon, 76

INDEX

importation of livestock, control of, 66
injurious plants, control of, 77
ivory, 151

JACKAL, 112
jackass, laughing, 72
jackdaw, 158

KANGAROO, 44, 64
kite, 19, 20
koala, 44

LABORATORY animals, 121, 172
ladybird, 75
lamprey, 55
Lantana, 79
lapwing, 22
leaf-beetle, 78
leather, 148
leopard, 35, 141, 176, 178, 181, 182, 187, 190, 191
lion, 29, 146, 175, 176, 177, 178, 179, 180, 181, 182, 186, 192
lizard, 56, 148, 162, 188
——, monitor, 148
llama, 110, 180
lobster, 23, 139
locust, 69, 142
lynx, 16, 176, 181

MAGPIE, 20, 188
malaria, 56, 168
mallard, 100
Malta fever, 171
mammals, marine, 48
manatee, 42
manure, 147
Maori, 41
marshlands, 18
marten, pine, 17, 18, 20, 62
martins, 160
martlet, 192
mastiff, 113
Mauritius, 39, 43
mealy bug, 75
medical services, 27
menageries, 173
mink, 95
minnow, top, 71

moa, 41
Mohammedanism, 35
Mollusca, 58
mongoose, 54, 66, 73, 140, 173, 190
monkey, 35, 142, 175, 187
——, pig-tailed, 123
moose, 15
mosquito, 56, 71, 167, 168, 170, 191
—— in rabbit control, 73
moth, gipsy, 75
——, silk, 101
—— for prickly pear control, 78
mouflon, 89
mouse, 54, 121, 178
——, field, 141
——, harvest, 22
——, house, 55, 122, 154
mule, 107
musk ox, 39
musquash, 95
mussel, 120, 138
mutton, 142

NATIONAL Parks, 31
natural history, 135
New Zealand, 41, 44, 57, 63, 72, 73, 76, 157
newt, 188
Notornis, 42

OCTOPUS, 139
opossum, 44
oryx, 174, 176
osprey, 19
ostrich, 96, 146, 176, 178
otter, sea, 49
otter-hunting, 129
owl, 173, 188
——, little, 60
——, tawny, 72
ox, Celtic, 88
——, domestic, 85, 86, 124, 148, 150, 153, 173, 176, 186
——, musk, 39
——, wild, 16, 46, 86, 175
oyster, 120, 139, 151

PARADISE, birds of, 149
—— fish, 120
parrot, 171
partridge, 22, 60, 119, 130

partridge, red-legged, 61
peacock, 119, 143, 149, 175
pearls, 120
Pegasus, 192
peregrine, 124
pets, 136, 173
pheasant, 54, 60, 119, 130, 131
pig, domestic, 85, 91, 124, 142, 165, 180
—— in New Zealand, 65
—— -sticking, 131
——, wild, 15
pigeon, carrier, 124
——, domestic, 117, 124, 157, 159, 186
——, passenger, 38
plague, bubonic, 155, 169
plover, little ringed, 25
polecat, 17, 18, 20, 71, 116
pollution, river, 24
ponies, 105
porcupine, 141
poultry, 96
prickly pear, 77
Primates, 122
psittacosis, 171
puff adder, 144
puffin, 143
puma, 180
python, 144, 148, 176

QUAGGA, 28, 29
quills, 149

RABBIT, 160
—— in Australia, 45, 54, 64, 73
—— in Gt. Britain, 62
—— in New Zealand, 64
—— in Spain, 73
—— control, 73
—— as pest, 129
—— in sport, 130
——, tame, 93
rabies, 170
rat, black, 54, 73, 141, 154, 162, 169
——, brown, 55, 122, 154
——, Gambian pouched, 154
——, giant cane, 141
——, musk, 94
——, tame, 93, 121
—— in Australia, 45

rat control in Zoo, 72
rattlesnake, 144, 187
raven, 188
redshank, 22
redstart, black, 22, 159
reindeer, 14, 61, 92, 124
rhinoceros, 42, 151, 176, 177
Rhyssa, 76
roach, 145
rodent, 154, 156
rooster, 190

ST. JOHN'S WORT, 78
salmon, 25, 66, 120, 145
—— flies, 149
sand grouse, 131
scabies, 165
schistosomes, 166
sea cow, Steller's, 42
sea cucumber, 139
seal, 177
——, common, 50
——, elephant, 50
——, fur, 48, 49
——, grey, 50
——, monk, 50
serpent, 173, 187
sewage, effects of, 24
—— farms, 24, 25
Seychelles Islands, 43
shark, 188
shark's teeth, 151
sharkskin, 148
sheep, domestic, 85, 86, 89, 124, 150, 166, 173, 174, 186, 190
—— -grazing, 14
——, introduced to Gt. Britain, 54
——, wild, 89, 90, 178
shellfish, 138
shells, 152
shrew, 173
Sierra Leone, 31, 35
silk, 101
silkworm, 152
skins, animal, 148
skylark, 66
slave-raiding, results of, 27
sleeping sickness, 169
smallpox, 172
snail, 166, 190
——, edible, 138
——, giant, 58, 77, 139

snake, grass, 188
——, West African house, 72
snakes, 144, 177, 190
snakeskin, 148
snipe-shooting, 131
solitaire, 41
sparrow, house-, 156
—— in Australia, 64
spider, 56, 135, 152, 189
squid, 139
squirrel, flying, 33
——, grey, 17, 59, 141
——, red, 17, 59
——, West African, 33
squirrels as food, 17, 141
stag, 128, 175, 192
—— -hunting, 134
starling in Australia, 64
—— in Canada, 62
—— in London, 158
stoat, 65, 66
stowaways, animal, 53, 56
string figures, 191
superstition, 188
swallow, 135, 160
swan, black, 64
——, mute, 118, 143
——, trumpeter, 39
swift, 160
swollen shoot in cocoa, 84

TAPEWORM, 165
teeth, animal, 150
termite, 84, 139, 153
tern, black, 24
terrapin, 144
thrush, 157
tick, 169
—— -bird, 153
tiger, 177, 179, 182, 188
tit, bearded, 22, 69
——, crested, 17
toad, 162, 188
——, giant, 73
tortoise, 144, 151, 190
——, giant, 43, 63, 144
totemism, 187
trepang, 139
trout, 25, 55, 120, 145
—— flies, 150

trypanosome, 169
tsetse fly, 27, 169, 190
turkey, 98, 187
turtle, 144, 151, 187
Tyrian purple dye, 151

UGANDA, 31

VEDALIA, 75
venom, snake, 147
vicuña, 110, 180
viper, Gaboon, 191
virus in cocoa trees, 84
vole, water, 94
vulture, 140, 159, 173, 190

WADING birds, 24, 25
wagtail, pied, 158
wallaby, 44
wapiti, 65
warbler, Dartford, 69
wasp, 57
water flea, 165, 166
waterbuck, 174
weasel, 66
weaver bird, 153, 161
weevil, alfalfa, 75
——, fern, 75
——, palm, 139
whale, 50
whelk, 138
White Horse, 188
winkle, 138
wisent, 37, 46, 47
wolf, 15, 114, 173, 185, 187, 188
wood-wasp, 76
wool, 150
worm, 190
wryneck, 22

YAK, 89, 150
yellow fever, 170

ZEBRA, Burchell's, 29
——, Grevy's, 29
zebu cattle, 68, 88